CONTENTS

CONTROLLING BIOTERROR: ASSESSING OUR NATION'S DRINKING WATER SECURITY

THURSDAY, SEPTEMBER 30, 2004

HOUSE OF REPRESENTATIVES,
COMMITTEE ON ENERGY AND COMMERCE,
SUBCOMMITTEE ON ENVIRONMENT
AND HAZARDOUS MATERIALS,
Washington, DC.

The subcommittee met, pursuant to notice, at 2:33, p.m., in room 2322, Rayburn House Office Building, Hon. Paul E. Gillmor (chairman) presiding.

Members present: Representatives Gillmor, Bass, Terry, Rogers, Solis, Pallone, Capps, and Stupak.

Staff present: Tom Hassenboehler, majority counsel; Mark Menezes, majority counsel; Jerry Couri, policy coordinator; Peter Kielty, legislative clerk; and Dick Frandsen, minority counsel.

Mr. GILLMOR. The subcommittee will now come to order and the Chair will recognize himself for the purposes of an opening statement.

Many people believe that our government's focus on protecting our infrastructure from terrorist attacks began after the attacks on the World Trade Center and the Pentagon. That assessment, I think, however, would be inaccurate.

It was over 50 years ago that the FBI was worried that an attack on our Nation's drinking water system could have devastating effects on the Nation's health and well-being. Unfortunately, it took the scare our Nation felt 3 years ago before Congress took action to fill in the legal gaps that prevented real preparedness from occurring.

I am proud that our committee took the bipartisan lead with the Public Health Security and Bioterrorism Response Act of 2002. This law took a major step forward in providing all drinking water systems with legal direction for safeguarding their product and their facilities.

This legislation not only recognized the need to protect our large urban centers or 15 percent of the community water systems that served 75 percent of the population, but also the need to look after smaller systems that did not have the resources but were every bit as vulnerable. Under this law, each community water system serving more than 3,300 individuals is required to conduct an assessment of the system's vulnerability to terrorist acts or other intentional acts to disrupt a safe and reliable drinking water supply. The Act also requires these systems to prepare or revise emergency response plans, incorporating the results of the vulnerability assessments and to do so no later than 6 months after completing them.

To pay for this, the Bioterrorism Act authorized funding to provide financial assistance to community water systems conducting those vulnerability assessments, preparing response plans and addressing the basic security enhancements.

And finally, the Act authorized the EPA to review methods by which terrorists or others could disrupt the provision of safe water supplies and identify methods for preventing, detecting, and responding to such disruptions.

These are all good things, but they are not—will not—deter us from conducting the kind of oversight to make sure the money spent is going to places where Congress intended; that water utilities are complying with the rules set forth, and that meaningful protections are occurring across our country.

In addition, we must know how the Federal apparatus that has since come into play, including the Department of Homeland Security, and subsequent Presidential directives are working in guiding drinking water protection from terrorism.

I also want to note one part of the bioterrorism law that I think was very important, and generated much discussion when we passed the law. In trying to provide for the collection of meaningful compliance information—and also discouraging the use of this information in inappropriate ways, Congress required water systems to certify to the EPA that they had conducted a vulnerability assessment and submitted a copy of the assessment to EPA. But, Congress also exempted the main contents of the vulnerability assessments from disclosure under the Freedom of Information Act.

In addition, the law directed EPA to develop protocols to protect the assessments from unauthorized disclosure and provides for civil and criminal penalties for inappropriate disclosure of information by government officials. I am interested in knowing how these protections are working.

And also, before closing my remarks, I want to thank our witnesses who are with us today: Mr. Benjamin Grumbles of the EPA, whom we have met with before; and Mr. John Stephenson, director of Natural Resources and Environment of GAO. We are very pleased to have you here, and we want you to know how much we value your input.

I would like to recognize the ranking member of our subcommittee, the gentlelady from California.

[The prepared statement of Hon. Paul Gillmor follows:]

PREPARED STATEMENT OF HON. PAUL GILLMOR, CHAIRMAN, SUBCOMMITTEE ON ENVIRONMENT AND HAZARDOUS MATERIALS

The Subcommittee will now come to order and the chair will recognize himself for 5 minutes for the purposes of delivering an opening statement.

Many people believe that our government's focus on protecting our infrastructure from terrorist attacks began after the terrible attacks on the World Trade Center and Pentagon that occurred on September 11, 2001. This assessment, however, would be false. Over 50 years ago, the Federal Bureau of Investigation was worried that an attack on our nation's drinking water system would have devastating effects on the nation's human and economic health and well-being. Unfortunately, it took the scare our nation felt three years ago before Congress took action to fill in the legal gaps that prevented real preparedness from occurring.

I am proud that our Committee took the bipartisan lead in fashioning the Public Health Security and Bio-terrorism Preparedness and Response Act of 2002. This law took a major step forward in providing all drinking water systems with legal directions for safeguarding their product and their facilities. I am particularly glad that this legislation not only recognized the obvious need to protect our large urban centers where 15 percent of all community water systems serve 75 percent of the population, but also the need to look after smaller systems that did not have the resources but were every bit, if not more vulnerable than their big city brethren.

This law should make a real difference in the future safety of drinking water systems. Under this law, each community water system serving more than 3,300 individuals is required to conduct an assessment of the system's vulnerability to terrorist attacks or other intentional acts to disrupt the provision of a safe and reliable drinking water supply. The Act also requires these systems to prepare or revise emergency response plans incorporating the results of the vulnerability assessments no later than 6 months after completing them. To fund all these items, the Bioterrorism Act authorized funding to provide financial assistance to community water systems conducting vulnerability assessments, preparing response plans, and addressing basic security enhancements and significant threats. Finally, the Act authorized EPA to review methods by which terrorists or others could disrupt the provision of safe water supplies, and identify methods for preventing, detecting, and responding to such disruptions.

These are all good things, but they should not deter us from conducting the kind of oversight that makes sure the money spent is going to the places Congress intended, the water utilities are complying with the rules set forth, and meaningful protections are occurring across our country. In addition, we must know how the Federal apparatus that has since come into play, including the Department of Homeland Security, and subsequent presidential directives are guiding drinking water protection from terrorism. This panel must not be deterred in appropriately exercising its jurisdiction over drinking water safety and supply. This is a charter that I take very seriously as Chairman and it is one that no committee in the House can or should do better.

I want, though, to make note of one part of the Bio-terrorism law that I find very important and that generated much discussion when we passed this law. In trying to carefully choreograph the dance between the collection of meaningful compliance information and discouraging the use of this information as a way to discover other places EPA might wish to regulate, Congress required water systems to certify to EPA that they have conducted a vulnerability assessment and submit a copy of the assessment to EPA, but Congress also exempted the main contents of the vulnerability assessments from disclosure under the Freedom of Information Act. In addition, the law directed EPA to develop protocols to protect the assessments from unauthorized disclosure, and provides for civil and criminal penalties for inappropriate disclosure of information by government officials. Forcing EPA to take sensitive information was only acceptable to many of us if serious protections were put into place on that material. I am interested in knowing how those protections are working.

Finally, before closing my remarks, I want to thank our witnesses who are here with us today. You know how much we value your input and look forward to your testimony and commentary.

With that, I want to recognize the Ranking Member of the Subcommittee, the gentle lady from California, Mrs. Solis, for 5 minutes for the purpose of delivering an opening statement.

Ms. SOLIS. Thank you, Mr. Chairman, and I thank you for holding this hearing today. This is a very important issue for many of us, the safety and security of our drinking water.

But I do have to say that I am a bit disappointed, because I believe it is a disservice to the members of the subcommittee that the Department of Homeland Security has failed to respond to our request to testify before this subcommittee. And I find it unfortunate that we are holding this hearing without the benefit of hearing from the Inspector General of the Environmental Protection Agency.

I understand we received a letter in our office, just today that was faxed over, informing us that they could not come but if we have any further questions, we could submit them. The inspector general has released numerous reports critiquing the status of security at facilities. And last September, the Inspector General found serious vulnerabilities in the EPA's action to combat contamination of our water supply as a result of terrorism.

The Inspector General highlighted that EPA had failed to sufficiently provide information about threats. Reports noted that as a result, major utilities were having problems identifying and prioritizing threats to our water supply.

I hope today that EPA will address the findings of the Inspector General because these vulnerabilities are unacceptable. We see the news in every day stories about drinking water being tainted by lead and rocket fuel and other contaminants.

When rocket fuel was found in milk, parents became outraged and questioned whether it was safe to serve their children milk. What would these same parents say if they knew about unaddressed vulnerabilities in drinking water? It is the responsibility of EPA to assure the public that not only have water utilities filed the necessary paperwork, but that the necessary upgrades have been made and that our drinking water is safe and secure.

The city of Los Angeles Department of Water and Power serves water and power to over 3.8 million users. In response to 9/11, Los Angeles devised a 5-year plan to increase daily sampling and test water quality. They installed security cameras, increased helicopter patrols and reinforced security barriers at water facilities.

These upgrades are being funded by an increase in consumer rates. I support the efforts of Los Angeles to secure its water facilities as much as possible. But I wonder where the EPA has been as Los Angeles makes these upgrades. What kind of role has EPA played in the development of the security policies of cities across the country? What kind of guidance have you given cities like Los Angeles. And why has the Inspector General time and time again revealed vulnerabilities in EPA's guidance? I hope we hear answers to these questions from EPA today.

Finally, I would also like to mention how unfortunate it is that another year has gone by and Congress still has not addressed chemical facility security.

There are more than 100 facilities nationwide, whose vulnerability puts at risk more than 1 million people each. As more than 60 editorials across the country have noted, Congress's inability or refusal to act to secure chemical facilities is a dereliction of duty. I sincerely hope this dereliction of duty isn't because my colleagues are afraid to challenge the chemical industry, the grave threat posed by chemical facilities is unneccessary—and I only hope that we will not be forced to regret this decision later.

Before I yield back the balance of my time, I would like to ask for unanimous consent that if the ranking member of the full committee, Mr. Dingell, and other and other members provide opening statements, that they be allowed to be inserted into the record.

Mr. GILLMOR. Without objection, it will be so ordered. All members will be able to submit opening statements to be inserted in the record.

Does the gentleman from Michigan have an opening statement?

Mr. ROGERS. I yield, Mr. Chairman.

Mr. GILLMOR. The gentlemen yields.

The gentlewoman from California.

Mrs. CAPPS. Thank you very much, Mr. Chairman, for holding this hearing.

The September 11 terrorist attacks on our country and the anthrax incidents that followed changed the way we looked at protecting our Nation's most critical infrastructure. They raised serious questions about our preparedness to respond to future catastrophic terrorist attacks, and they made Americans concerned about all aspects of our safety, including our drinking water supply.

The President raised drinking water concerns in his 2002 State of the Union Address. He stated that U.S. Forces in Afghanistan found diagrams of U.S. Public water utilities and that we are under "continuing and immediate threats of future attacks. A successful terrorist attack on a public water system would be devastating."

In addition, it would further damage public confidence and safe and reliability supplies of drinking water, so we need this afternoon to assess whether the administration is pursuing an effective strategy to prevent or to respond to such attacks. Many experts are concerned, and for good reason.

In fact, the EPA's Inspector General has conducted preliminary research on how well we have evaluated water system security activities. As you know, EPA has responsibility over a safe-guarded water supply. In 2002, Congress passed a Bioterrorism Act, and this bill required the water utilities to assess vulnerabilities based on threat information provided by EPA and then submit these vulnerability assessments to EPA for their review. Vulnerability assessments are a necessary tool for drinking water utilities to evaluate and identify their vulnerabilities.

But vulnerability assessments alone don't protect us from the threats. They only detect them. And according to the Inspector General, EPA may not be taking the necessary steps to maintain a safe and reliable drinking water supply. For example, the IG has reported that due to limited threat information provided by EPA, the utilities design their assessments around pre-September 11 threats.

Based on interviews with key stakeholders, the IG went on to include "we believe that vulnerability assessments submitted may emphasize traditional less consequential and less costly threats such as vandalism or disgruntled employees."

This is certainly not what we are talking about post-9/11. How can we expect to adequately assess the specific shortcomings of our public water systems, much less implement protective measures, without an accurate evaluation of realistic threats. The world changed on September 11. It changed the way we need to detect and protect ourselves from threat.

Yet if the IG's reports is accurate that water utilities were not provided with updated and accurate threat information, then the water utilities may be defending our water supply from yesterday's disgruntled employees instead of today's enemies. Merely completing the vulnerability assessments are not going to reduce these unacceptable security risks.

We have to also take corrective actions to reduce them. While I am very disappointed the IG is not able to join us to discuss these important issues, I certainly hope our witnesses will address his concerns. In his absence, I ask unanimous consent to insert two evaluation reports from the Office of the Inspector General dated September 11, 2003 and September 24, 2003 into the record, Mr. Chairman.

Mr. GILLMOR. Without objection.

[The information referred to follows:]

OFFICE OF INSPECTOR GENERAL

Catalyst for Improving the Environment

Evaluation Report

EPA Needs a Better Strategy to Measure Changes in the Security of the Nation's Water Infrastructure

Report No. 2003-M-00016

September 11, 2003

Report Contributors: Erin Barnes-Weaver
Eric Hanger
Jeffrey Harris
Fredrick Light
Ricardo Martinez
Erin Mastrangelo

Abbreviations

DHS	Department of Homeland Security
EPA	Environmental Protection Agency
ISAC	Information Sharing and Analysis Center
VA	Vulnerability Assessment

UNITED STATES ENVIRONMENTAL PROTECTION AGENCY
WASHINGTON, D.C. 20460

THE INSPECTOR GENERAL

MEMORANDUM

SUBJECT: Final Evaluation Report: EPA Needs a Better Strategy to Measure Changes in the Security of the Nation's Water Infrastructure
Report No. 2003-M-00016

FROM: Nikki L. Tinsley

TO: G. Tracy Mehan, III
Assistant Administrator for Office of Water

As part of our ongoing evaluation of Environmental Protection Agency (EPA) activities to enhance the security of the Nation's water supply, we noted an issue that requires your attention. Specifically, we suggest that EPA develop specific measurable performance indicators of water security activities. We propose this action because, during our preliminary research,[1] we obtained information that indicates EPA has neither:

- Articulated measurable goals for EPA's water security efforts; nor
- Obtained or analyzed data to develop a baseline for water security.

It is important for EPA to develop measures to monitor the security of our Nation's water supply and to ensure Federal funds are not spent without clear goals or expectations. To effectively perform its lead agency responsibilities, EPA needs to collect and analyze data that depicts the changes in security levels at water utilities.

Our observations and suggestions are based on information obtained from our interviews with water security experts, water utility officials, and EPA headquarters and regional representatives; attendance at water vulnerability assessment training; and a review of EPA's *Strategic Plan for Homeland Security*, dated September 2002. We are performing our evaluation in accordance with *Government Auditing Standards,* issued by the Comptroller General of the United States.

EPA Designated as Lead Agency for Water Security

The Nation's water supply is one of our most vital natural resources. Potential threats to this resource include contamination with biological, chemical, or radiological agents, or destruction of physical infrastructure. The water supply is also dependent on other critical infrastructures,

---

[1]The EPA Office of Inspector General is conducting preliminary research on an evaluation of water system security activities in support of the Agency's *Strategic Plan for Homeland Security*.

such as energy and transportation. Presidential Decision Directive 63, issued in May 1998, designated EPA as the lead agency responsible for the water sector and for accomplishing the following functions:

- Establish and maintain channels of communication with all private and public entities having an infrastructure assurance interest in the sector;
- Facilitate the selection of a Sector Infrastructure Assurance Coordinator;
- Assist the Sector Coordinator in establishing and operating an effective information-sharing program;
- Draft new legislation and regulations, as required, and propose the use of Federal incentives to facilitate private investment in assurance programs, if appropriate;
- Promote infrastructure assurance education and training, to include advocating use of best practices, within the sector;
- Assist in developing plans for prevention (long-term reduction of vulnerabilities and short-term defensive actions), mitigation, restoration, and reconstitution; and
- Coordinate, in support of the Federal Response Plan, as amended, management of the consequences of a successful infrastructure attack and prepare for various contingent attacks through participation in training and exercise programs.

EPA's Efforts to Improve Water Security

To better execute its responsibilities after the terrorist attacks on September 11, 2001, EPA developed a strategy for improving the security of water utilities. EPA's Strategic Plan focuses on preparedness and prevention, assisting those responsible for critical infrastructures in assessing and reducing vulnerabilities and maximizing their response capabilities. Also, EPA intends to develop technologies to improve the Nation's critical infrastructure and key responders' abilities to detect and monitor environmental threats. Through this work, EPA plans to "significantly improve the Nation's overall capacity to protect critical infrastructure from terrorist attacks." EPA will rely on relationships with water utilities, water-related governmental entities, and associations to assist utilities, and the Agency has already taken the following steps:

- Facilitated the development of vulnerability assessment methodology and training;
- Provided threat guidance to utilities to help them conduct vulnerability assessments and identify possible threats to critical assets;
- Provided financial assistance to large drinking water systems to conduct one-time vulnerability assessments;
- Funded research on technology development and verification; and
- Facilitated the development of a secure Information Sharing and Analysis Center for the water utility sector (Water-ISAC) to exchange threat/incident information.

EPA's Strategic Plan is organized into four mission-critical areas, the first being "Critical Infrastructure Protection." This area states that EPA will work with the States, tribes, drinking water utilities, and other partners to enhance the security of water utilities. The Strategic Plan articulates both the tactics to execute the Plan as well as the anticipated results. The following table illustrates selected tactics and results.

Selected Water Infrastructure Protection Tactics	Anticipated Results
Technical Assistance and Grants EPA will provide tools, training, and technical assistance to assist water utilities in conducting vulnerability assessments, implementing security improvements, and effectively responding to terrorist events. EPA provided grants to large drinking water utilities for vulnerability assessments, security enhancement designs, and/or emergency response plans.	By the end of fiscal 2003, all water utility managers will have access to basic information to understand potential water threats, and basic tools to identify security needs. By 2005, unacceptable security risks at water utilities across the country will be significantly reduced through completion of appropriate vulnerability assessments, design of security enhancement plans, development of emergency response plans, and implementation of security enhancements.
Terrorism Methods and Prevention Techniques EPA will work with the Department of Homeland Security (DHS), other Federal agencies, universities, and the private sector to: – Solicit and review methods to prevent, detect, and respond to chemical, biological, and radiological contaminants that could be intentionally introduced in drinking water systems; – Review methods and means by which terrorists could disrupt the supply of safe drinking water or take other intentional actions against water collection, pretreatment, treatment, storage, and distribution facilities; and – Review methods and means by which alternative supplies of drinking water could be provided in the event of a disruption.	Starting in fiscal 2003, water utilities, key response agencies, and policymakers will have improved information and knowledge to make timely and effective analytical and technological decisions to enhance security, detect contamination, and respond to incidents.
Security Practices EPA will work to implement water security practices in ongoing water utility operations. EPA will also work to build security concerns into ongoing review systems (e.g., sanitary survey, capacity development, operator certification, and treatment optimization program for drinking water systems).	Beginning in fiscal 2003, water utilities will incorporate security measures as a standard aspect of day-to-day operations and EPA, States, and tribes will review security measures at water utilities on a continuous basis. Through ongoing practice and review, water utility managers and employees will optimize security measures.

Selected Water Infrastructure Protection Tactics	Anticipated Results
Communication with Utilities EPA will work with other government agencies, utility organizations, and water utilities to establish formal communication mechanisms to facilitate the timely and effective exchange of information on water utility security threats and incidents.	Starting in fiscal 2003, water utilities, law enforcement agencies, and State and Federal response and prevention programs will have timely and accurate security threat information and incident analysis to make effective decisions for water security preparedness and response.
Coordination with First Responders EPA will work in coordination with DHS to foster coordination among Federal, State, tribal, and local emergency responders, health agencies, environmental and health laboratories, the medical community, and the law enforcement community at all levels (Federal, State, and local) concerning response to potential terrorist actions against water utilities. This will be achieved through training and support of simulations and emergency response exercises.	In the majority of water security incident responses and exercises, the decision-making and communication structures of response agencies will function smoothly (without critical errors).
Coordination with Other Critical Infrastructures EPA will work with other critical infrastructure sectors to further understand and reduce the impact to water utilities of terrorist attacks on related infrastructures as well as the impacts of attacks on water utilities to other critical infrastructures.	Water sector vulnerabilities and impacts resulting from attacks on other critical infrastructure sectors will be reduced and vice versa.

EPA Has Not Articulated Measurable Goals for Water Security

The Office of Water has not outlined how resources, activities, and outputs will achieve the water security program's goals. EPA's Strategic Plan lacks fundamental components, such as measurable performance results and information and analysis, to ensure the greatest practicable reductions in risks to the critical water sector infrastructure. We based our observations on key program management practices consistent with the President's Management Agenda and the Government Performance and Results Act.[2]

[2]The EPA Office of Inspector General has compiled these program management principles in *Assessing Organizational Systems: A User's Guide,* OA/OPE-5, November 5, 2002. The *Guide* identifies seven areas requiring management attention for a successful program. The seven areas consist of Leadership, Strategic Planning, Customer/Stakeholder and Market Focus, Information and Analysis, Human Capital, Process Management, and Performance Results.

Because EPA lacks the indicators that define the baseline for water security, EPA cannot monitor program performance against goals. Without this baseline, EPA cannot determine whether its strategy resulted in improved water security. Officials we interviewed at the Office of Management and Budget and DHS also endorsed the need for EPA to establish performance indicators to determine the effectiveness of its water security activities.

In the absence of specific measurable water security goals, the Agency has focused on compliance with legislative requirements. For example, EPA states, "By 2005, unacceptable security risks at water utilities across the country will be significantly reduced through completion of appropriate vulnerability assessments; design of security enhancement plans; development of emergency response plans; and implementation of security enhancements." The completion of these documents, however, does not equate to the outcome of reducing unacceptable security risks.

EPA Needs to Obtain and Analyze Information to Measure Changes in Security

For EPA to develop performance indicators that measure changes in security, EPA needs to collect and analyze information from water utilities. However, the Agency believes that it: (1) lacks the authority to ask for information directly from water utilities or utilize information in the Water-ISAC that would show changes in security levels, and (2) cannot analyze information in vulnerability assessments because that would violate the Public Health Security and Bioterrorism Preparedness and Response Act (Bioterrorism Act).

However, we believe that EPA does in fact have the authority, as well as the responsibility, to collect and analyze necessary information from these sources. Information obtained anonymously, through the Water-ISAC, or through coordination with DHS, would provide EPA with data to benchmark changes in security levels. EPA needs to review the vulnerability assessments to identify and prioritize threats to water utilities. Subsection 6(A) of the Bioterrorism Act appears to contemplate review and analysis of vulnerability assessments by duly authorized Agency officials. Further, additional support for vulnerability assessment analysis is found in the legislative history of the Bioterrorism Act. Specifically, congressional members showed bipartisan support for EPA's review of the vulnerability assessments to develop plans to protect drinking water supplies, and noted the Agency's discretion to review the assessments and make recommendations to improve water security.[3]

- Representative Frank Pallone, Jr. (D-NJ) stated that EPA's review of the findings of the vulnerability assessments would "help the government understand the threats to our water systems and develop plans to protect our safe drinking water supply."

- Representative Michael Bilirakis (R-FL) stated that EPA could use the assessments "to address the threat of terrorism and for any other lawful purpose."

---

[3]Floor Action on H.R. 3448, the Public Health Security and Bioterrorism Preparedness and Response Act of 2002, in the Senate on May 23, 2002, and in the House on May 22, 2002.

- Senator James Jeffords (I-VT) stated that "there is not a restriction on EPA's discussing the content of the assessments with persons who may benefit from information about the security of our Nation's water supply, such as state and local officials, nor is there restriction intended by this bill upon a water system's voluntarily sharing information with other systems, emergency responders or communities. Our attempt to provide a safeguard against broad disclosure of sensitive information does not lead us to conclude that our citizens should not have the information they need to protect and inform themselves."

EPA could also utilize information from DHS to develop performance indicators that illustrate changes to security at critical infrastructures. An official at DHS stated that the Department is developing performance indicators for other critical infrastructures, such as the chemical industry. According to *The National Strategy for the Physical Protection of Critical Infrastructure and Key Assets* (February 2003), DHS coordinates with Federal agencies to assess threats to critical infrastructure and evaluate preparedness. The National Strategy places responsibility with DHS for gathering threat and vulnerability information. However, EPA has only recently allowed DHS to review drinking water utilities' vulnerability assessments.

Suggestions

We suggest that EPA:

(1) Develop specific, measurable goals, objectives, and performance indicators for its water security programs; and

(2) Utilize available sources of information to collect and analyze data to develop a baseline for water security.

Agency Comments and Office of Inspector General Evaluation

In its response to our draft report, EPA agreed with our assessment that the Agency's *Strategic Plan for Homeland Security* lacks clearly defined performance measures for critical water infrastructure protection activities. EPA indicated that it will be actively involved in developing more outcome-focused performance indicators in its revised strategic plan.

EPA also agreed with our assessment that it needs to develop a baseline for water security. EPA said it will analyze a sample of vulnerability assessments to develop the baseline. EPA also plans to use this information to assist in identifying tools that need to be developed, and to research priorities and appropriate security enhancements for water utilities. EPA indicated it will share this analysis with Congress and the Department of Homeland Security.

We commend EPA's commitment to create outcome-focused performance measures and a baseline to measure and monitor changes in water security. However, the necessary information required for these initiatives may not reside solely in the water system vulnerability assessments. As the lead agency for water security, we suggest that EPA partner with the utilities and collect and analyze information from additional sources such as the Water-ISAC and through coordination with DHS so that the appropriate actions could be taken to enhance the security of the Nation's water infrastructure.

The full Agency response is provided in Appendix A.

- - -

If you or your staff have any questions regarding this report, please call me at (202) 566-0847 or Kwai Chan, Assistant Inspector General for Program Evaluation at (202) 566-0828.

Appendix A

Agency Response

August 22, 2003

MEMORANDUM

SUBJECT: Response to OIG Concerns regarding "EPA Needs a Better Strategy to Measure Changes in Utilities Water Security"
DRAFT: Report No. 2003-M-00016

FROM: G. Tracy Mehan, III /signed/
Assistant Administrator

TO: Jeffrey K. Harris
Director for Program Evaluation, Cross-Media Issues
Office of Inspector General

I am responding to the two principal issues and concerns that were stated in your July 23, 2003, memorandum/report to me. Your preliminary research on performance measures of water security activities indicates that EPA has neither:

- articulated measurable goals for EPA's water security efforts; nor
- obtained or analyzed data to develop a baseline for water security.

I agree with your assessment that the Agency's *Strategic Plan for Homeland Security*, dated September 2002, lacks clearly defined performance measures for critical water infrastructure protection activities. I believe this document should be considered a blueprint of near-term (2002 and 2003) homeland security activities rather than a strategic plan in the strict sense of that term. Attachment 1 highlights our significant accomplishments under this near-term plan. I am a strong proponent of EPA's Office of Homeland Security's recently-announced initiative to develop a revised strategic plan. This exercise may serve as the opportunity to identify longer-term goals, objectives, and measurable performance indicators for all EPA programs involved in high priority activities to protect public health and the environment from terrorist and other intentional acts. I assure you that my staff will be very active participants in this endeavor and will devote significant attention to the challenge of developing outcome-focused performance measures. For the FY 2005-2010 Agency-wide strategic plan, I assigned my Senior Advisor for Water Policy to work with all water programs to shift from a predominance of output-oriented performance measures to ones that will demonstrate more direct results and benefits to public health and environmental protection. We made considerable progress and will continue to make improvements in all water programs including critical water infrastructure protection.

In addition to the limited performance measures in the *2002 Strategic Plan for Homeland Security*, the Agency's FYs 03 and 04 annual performance plans have contained annual performance goals and measures. (Attachment 2) Because these measures are indeed outputs not indicators of outcomes, I and my staff would welcome any specific recommendations and direct assistance from you as we proceed to formulate water security program goals and appropriate performance measures.

With respect to your second issue/concern on obtaining and analyzing data to develop a baseline for water security, I want to clarify the Office of Water's position on examining vulnerability assessments (VAs) submitted by some 9,000 community water systems as required by the Bioterrorism Act of 2002. I personally identified to the few members of the Water Protection Task Force who have been designated by me to have access to the VAs three specific purposes for reviewing and analyzing them. The first is that **all** VAs should be reviewed to determine both compliance with the statutory requirements for submission/ certification of completion and necessary enforcement action. Second, a **representative sample** (based on the total number of systems in each size of systems as stipulated in the statute) of VAs should be examined for: a) compliance (and subsequent enforcement, if any) with the statutory requirement that the VA address all applicable parts of a system (e.g., pipes, physical barriers, treatment, etc.) and b) determination that drinking water systems used a "reasoned process" (e.g., a tool like RAM-W) to evaluate their vulnerabilities. Third, **aggregated data** are to be compiled and analyzed to assist in identifying tools that need to be developed, research priorities and appropriate security enhancements. In addition, this aggregated information will help us develop a baseline for water security, which is consistent with your suggestion as well as OW's emphasis on improving performance indicators and measures for all its programs. Besides using this aggregated data to inform and guide our future actions, we intend to share this information (mindful of the restrictions imposed by the Bioterrorism Act of 2002) with Congress and other decision makers, e.g., the Department of Homeland Security.

I appreciate the opportunity to respond to your draft report. Should you have any questions or need additional information, please contact Michael Mason, the Office of Water's liaison to OIG, on 564-0572.

Attachments

Attachment 1

ACCOMPLISHMENTS OF THE WATER PROTECTION TASK FORCE UNDER EPA'S HOMELAND SECURITY STRATEGIC PLAN
7/24/03

EPA has and is continuing to support a number of activities to improve security of both drinking water and wastewater utilities using approximately $90 million and $23 million appropriated in Fiscal Years '02 and '03, respectively. Examples of these activities include:

- Developed vulnerability assessment tools for both drinking water and wastewater utilities and supported extensive training for thousands of utility operators.
- Provided more than $50 million in grants to more than 400 of the Nation's largest drinking water systems to undertake vulnerability assessments and do related security planning. Visited 30 of the largest cities to discuss water security. Received 463 of the 466 vulnerability assessments required to be submitted to EPA in March 2003.
- Provided more than $24 million to the States and non-profit organizations to provide training and technical assistance to small and medium water utilities on vulnerability assessments.
- Developed and distributed *Baseline Threat Information for Vulnerability Assessments of Community Water Systems*.
- Met statutory deadline to implement protocol to protect vulnerability assessments.
- Supported establishment of a state-of-the-art, secure information sharing system (the WaterISAC) to share up-to-date threat and incident information between the intelligence community and the water sector. Provided several water-specific advisories.
- Developed guidelines on what actions utilities should take under DHS-specified threat levels.
- Developed and began testing/distribution of *Riverspill* and *Pipeline Net* models to determine fate and transport of contaminants in both source water and drinking water systems.
- Supported establishment of a water security emphasis for the Environmental Technology Verification Program and the WATERS test site to evaluate water security technologies.
- Collaborated with ORD in development of the Water Security Research and Technology Development Action Plan, currently under Review by the National Academy of Science.
- Developed initial guidance on emergency response notification protocols. Currently developing detailed guidance on revising emergency response plans to meet Bioterrorism Act requirements, as well as a protocol to respond to a drinking water contamination event.
- More information on EPA's Water Protection Task Force water security program can be found at: www.epa.gov/safewater/security

Attachment 2

Annual Performance Goals and Measures for Critical Water Infrastructure Protection
(Drinking Water and Wastewater Utilities)

Drinking Water

FY 03:

Annual Performance Goal

Enhance public health protection by securing the Nation's critical water infrastructure through support for counter-terrorism preparedness.

Percent of the population and the number of community water systems - - serving 100,000 or more people - - that have certified the completion of their vulnerability assessment and submitted a copy to EPA.

Percent of the population and the number of community water systems - - serving 100,000 or more people - - that have certified the completion of the preparation or revision of their emergency response plan..

FY 04:

Annual Performance Goal

Enhance public health protection by securing the Nation's critical water infrastructure through support for counter-terrorism preparedness.

Annual Performance Measures

Percent of the population and the number of community water systems - - serving more than 50,000 but less than 100,000 people - - that have certified the completion of their vulnerability assessment and submitted a copy to EPA.

Percent of the population and the number of community water systems - - serving more than 50,000 but less than 100,000 people - - that have certified the completion of the preparation or revision of their emergency response plan.

Percent of the population and the number of community water systems - - serving more than 3,300 but less than 50,000 people - - that have certified the completion of their vulnerability assessment and submitted a copy to EPA.

Annual Performance Goals and Measures for Critical Water Infrastructure Protection

Wastewater

FY 03:

Annual Performance Goal

Enhance public health and environmental protection by securing the Nation's critical water infrastructure through support for counter-terrorism preparedness including system operator training.

Annual Performance Measure

Percent of the population and the number of large and medium size (10,001 and larger) of Publicly Owned Treatment Works (POTWs) that have been taken for homeland security preparedness.

FY 03:

Annual Performance Goal

Enhance public health and environmental protection by securing the Nation's critical water infrastructure through support for counter-terrorism preparedness including system operator training.

Annual Performance Measure

Percent of the population and the number of large and medium size (10,001 and larger) of Publicly Owned Treatment Works (POTWs) that have been taken for homeland security preparedness.

Appendix B

Distribution

Acting Administrator
Associate Administrator for Congressional and Intergovernmental Relations
Acting Associate Administrator, Office of Public Affairs
Assistant Administrator, Office of Water
Audit Followup Coordinator, Office of Water
Director, Office of Ground Water and Drinking Water
Chair, Water Protection Task Force
General Counsel
Director, Office of Homeland Security

OFFICE OF INSPECTOR GENERAL

Catalyst for Improving the Environment

Evaluation Report

EPA Needs to Assess the Quality of Vulnerability Assessments Related to the Security of the Nation's Water Supply

Report No. 2003-M-00013

September 24, 2003

Report Contributors: Erin Barnes-Weaver
Eric Hanger
Fred Light
Ricardo Martinez
Erin Mastrangelo

Abbreviations

AWWARF	American Water Works Association Research Foundation
CDC	Centers for Disease Control and Prevention
EPA	Environmental Protection Agency
RAM-W	Risk Assessment Methodology for Water

UNITED STATES ENVIRONMENTAL PROTECTION AGENCY
WASHINGTON, D.C. 20460

OFFICE OF
INSPECTOR GENERAL

September 24, 2003

MEMORANDUM

SUBJECT: EPA Needs to Assess the Quality of Vulnerability Assessments Related to the Security of the Nation's Water Supply
Report No. 2003-M-00013

FROM: Jeffrey K. Harris /s/
Director for Program Evaluation, Cross-Media Issues

TO: Tracy Mehan
Assistant Administrator for Office of Water

In connection with our ongoing evaluation of the Environmental Protection Agency's (EPA's) activities to enhance the security of the Nation's water supply, we noted an issue that requires your immediate attention. Specifically, we believe EPA should promptly analyze the vulnerability assessments submitted by large utilities pursuant to the Public Health Security and Bioterrorism Preparedness and Response Act of 2002 ("Bioterrorism Act") to determine whether the assessments adequately and comprehensively address terrorist threats.

We propose this action because, during our preliminary research,[1] we obtained information that suggests that problems may exist in:

- Identifying and prioritizing specific threats – particularly terrorist scenarios; and
- Assessing the full breadth of a water system's infrastructure – particularly its distribution system.

It is important that EPA promptly implement improvements to the vulnerability assessment process. According to an EPA official, although approximately 400 large utilities already submitted their vulnerability assessments, thousands of additional assessments are due from medium-sized water systems before the end of 2003 and from small-sized utilities by mid-2004.

[1]The EPA Office of Inspector General is conducting preliminary research on an evaluation of water system security activities in support of the Agency's Strategic Plan for Homeland Security.

Therefore, we determined that our observations were significant enough to report to you at this time because of the time-critical nature of the issues discussed below. The Bioterrorism Act authorized $160 million for fiscal year 2002 – and such sums as may be necessary for fiscal years 2003 through 2005 – to fund water security activities, including the vulnerability assessments, and Congress may base future funding decisions on those assessments.

Our observations and suggestions are based on information obtained from our interviews with water security experts, water utility officials, and EPA headquarters and regional representatives; attendance at water vulnerability assessment training; and a review of vulnerability assessment tools, methodologies, and related documents. We are performing our evaluation in accordance with *Government Auditing Standards,* issued by the Comptroller General of the United States.

Vulnerability Assessments Provide Foundation for Emergency Response

The nation's water supply is one of our most vital natural resources. Potential threats to this resource include contamination with biological, chemical, or radiological agents, or destruction of physical infrastructure. Presidential Decision Directive 63, issued in May 1998, designated EPA as the lead agency for assuring the protection of the nation's water infrastructure. The terrorist attacks on September 11, 2001 ("9/11") resulted in passage of the Bioterrorism Act and its requirement that water utilities submit vulnerability assessments to EPA. EPA's strategy for improving water security relies on water utilities to conduct vulnerability assessments, develop or modify emergency response plans, and institute security enhancements. EPA facilitates these actions by developing assessment tools and training, compiling a single threat summary, and providing financial assistance directly to large drinking water systems to conduct vulnerability assessments and to States for medium- and small-sized utilities.[2]

Figure 1 illustrates that vulnerability assessments serve as the foundation for emergency response plans and future security enhancements implemented by water utilities. EPA's November 2002 *Vulnerability Assessment Factsheet* notes that vulnerability assessments help water systems evaluate susceptibility to potential threats and design response plans and corrective actions to lessen the risk of serious consequences. EPA's *Factsheet* further states that an effective vulnerability assessment serves as a guide to the water utility by providing a prioritized plan for security upgrades, modifications of operational procedures, and/or policy changes to reduce risks to a utility's critical assets. A water security expert at Sandia National Laboratory[3] said that utilities use vulnerability assessments to help determine how well water systems detect security problems and stop or delay undesired events, as well as measure response capabilities. In

---

[2]Following the terrorist attacks of 9/11, EPA received supplemental fiscal 2002 funding of $89 million to improve the safety and security of the Nation's water supply. EPA used $53 million of that funding to provide grants to the largest water utilities (those that serve 100,000 or more people) to assist them in conducting vulnerability assessments. EPA also provided $21 million in grants to assist States in improving drinking water security for medium utilities (serving between 50,000 and 99,999 people) and small utilities (serving between 3,300 and 49,999 people).

[3]A Government-owned facility operated by a contractor for the U.S. Department of Energy's National Nuclear Security Administration.

addition to water utilities, vulnerability assessments are routinely used to develop response plans to address threats to chemical facilities, computer systems, nuclear weapons facilities, the electrical power industry, and wastewater treatment plants. Figure 1 also illustrates EPA's efforts in the water security area and shows the Agency's primary role in providing vulnerability and threat assessment assistance.

Figure 1: Water Security Concept Model

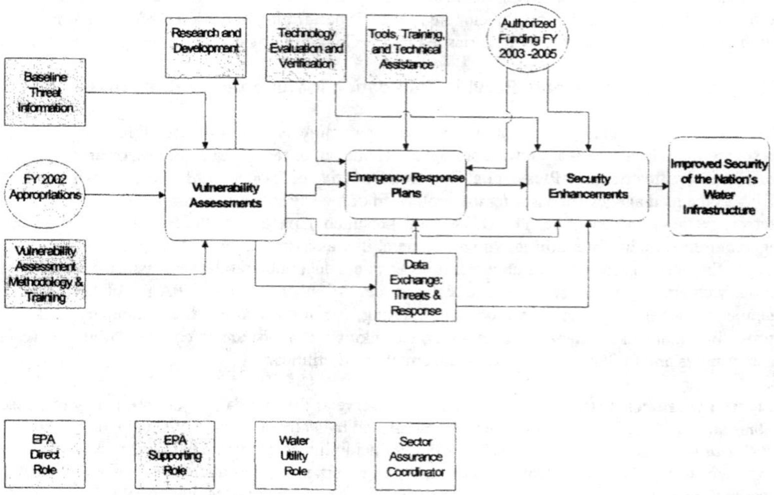

The Bioterrorism Act required that utilities serving a population greater than 3,300 persons conduct and submit their vulnerability assessments to EPA according to deadlines based on a utility's size.[4] Water utilities may conduct their assessments using one of several different methodologies. EPA provided funding to Sandia National Laboratory and the American Water Works Association Research Foundation (AWWARF) to develop training on the Risk Assessment Methodology for Water (RAM-W). RAM-W is one tool utilities can use to

---

[4]Water utilities serving 100,000 or more users had to submit their assessments by March 31, 2003; mid-sized utilities serving between 50,000 and 99,999 users must submit their assessments by December 31, 2003; and small utilities serving between 3,300 and 49,999 users must submit their assessments by June 30, 2004.

systematically assess vulnerabilities to terrorist and other intentional attacks. The RAM-W program stems from a vulnerability assessment methodology initially developed by Sandia to support the national nuclear security mission and from Sandia's involvement in the development of a risk assessment approach for dams. Sandia provided RAM-W training workshops under an interagency agreement with EPA. While EPA focused its efforts on the development of RAM-W, water utilities have other methodologies available to assist them in conducting their vulnerability assessments. EPA provided assistance to the Association of Metropolitan Sewerage Agencies, the Association of State Drinking Water Administrators, and the National Rural Water Association to develop similar tools to help medium and small utilities assess threats to their water systems.

Regardless of the methodology used, the Bioterrorism Act identified six elements that water utilities must address in their assessments of vulnerabilities to a terrorist attack or other acts intended to substantially disrupt the ability to provide a safe and reliable supply of drinking water:

(1) Pipes and constructed conveyances.
(2) Physical barriers.
(3) Collection; pretreatment; and treatment, storage, and distribution systems.
(4) Electronic or computer systems.
(5) Use, storage, and handling of chemicals.
(6) System operation and maintenance.

EPA issued guidance to utilities interpreting the Bioterrorism Act's six elements in the Agency's November 2002 *Vulnerability Assessment Factsheet.* While EPA did not specify a particular format or methodology for the vulnerability assessments, EPA emphasized that the following guidance applies to the vulnerability assessments conducted by all water utilities regardless of the size of the population served:

(1) Characterization of the water system, including its mission and objectives.
(2) Identification and prioritization of adverse consequences to avoid.
(3) Determination of critical assets that might be subject to malevolent acts that could result in undesired consequences.
(4) Assessment of the likelihood (qualitative probability) of such malevolent acts from adversaries.
(5) Evaluation of existing countermeasures.
(6) Analysis of current risk and development of a prioritized plan for risk reduction.

Vulnerability Assessments May Not Necessarily Address Terrorist Threats

The Bioterrorism Act requires community water systems to prepare for and assess vulnerabilities to terrorist and other intentional acts. However, based on our interviews, we believe that vulnerability assessments submitted may emphasize traditional, less consequential, and less costly threats, such as vandalism or disgruntled employees. Therefore, vulnerability assessments may not necessarily address terrorist scenarios or the events of 9/11 that motivated passage of the Bioterrorism Act.

The assessment of vulnerabilities is a threat-driven exercise where the design of response actions are dependent upon the credibility of the defined threat. Neither the Bioterrorism Act nor EPA identified a minimum threat level against which water utilities should assess their vulnerabilities. Water security experts view understanding the threat as the driver to vulnerability assessment methodologies. However, EPA provided limited threat information that resulted in utility managers having to determine threats and response actions themselves. The RAM-W methodology instructed managers to define their system-specific threat by considering their own operational, legal, and financial limitations against the threat information provided by local intelligence sources.

Water security experts we interviewed stated that EPA did not provide adequate threat information. Officials at Sandia National Laboratory stated that EPA's threat guidance missed the mark because EPA did not set a minimum threat level against which utilities needed to assess their vulnerabilities. One AWWARF official found EPA's threat guidance too general and believed it lacked information utilities could act upon. For example, the document left responsibility to the utilities in defining subjective terms such as "reasonable protective measures." The AWWARF official further stated that EPA made no effort to provide credible threat information to utilities who needed it. The official said that although the Centers for Disease Control and Prevention (CDC) worked on compiling a list of potential contaminants, neither EPA nor CDC distributed this information to utilities. Although EPA incorporated the CDC information into the Agency's State of Knowledge report on contaminant threats, EPA officials considered that report to be too sensitive to share with decision-makers, including utility managers and congressional staff. Consequently, the AWWARF official noted that the Bioterrorism Act tasks utilities with conducting vulnerability assessments without proper credible threat information from EPA.

In the absence of credible threat information from EPA, water utility staff decided for themselves what threats to include in their vulnerability assessments. For example, one water security expert, contracted to conduct vulnerability assessments for many large water systems, said that, despite the RAM-W training provided after 9/11, water utilities focus on vandals, criminals, and disgruntled employees in their vulnerability assessments. The contractor further stated that EPA has not provided utilities the intelligence data or threat information required to justify the security upgrades necessary to defend against terrorism.

While the terrorist attacks of 9/11 and the subsequent passage of the Bioterrorism Act served as the catalyst for the vulnerability assessments, limited threat information provided by EPA resulted in utilities subjectively designing their assessments around pre-9/11 threats. All of the utilities and contractors we interviewed used the RAM-W methodology to complete their vulnerability assessments. After filtering threat information through the RAM-W methodology, most of the water security experts we interviewed who were familiar with vulnerability assessments concluded that the only threats utilities could realistically address were those they encountered before 9/11. One utility representative we interviewed said that the contractor they hired to conduct their vulnerability assessment discouraged them from addressing higher threat levels like terrorism.

Assessment Guidance Does Not Emphasize Unique Water System Vulnerabilities

Neither EPA nor the vulnerability assessment methodologies provided threat guidance that identified the most vulnerable components unique to water systems. The lack of clear guidance on what components to focus on resulted in utility managers deciding for themselves whether to emphasize the vulnerabilities of components, such as distribution systems. This results in inconsistent assessments and response actions, and may prevent EPA from ensuring future improvements to water security.

Many experts view water distribution systems as the most susceptible to terrorist attack. Such experts included the President of the Association of Metropolitan Water Agencies, who concluded that water distribution systems remain the most vulnerable to terrorist threats and could spread highly concentrated amounts of poison to a few thousand homes or businesses. The Chair of the National Academy of Sciences' Water Science and Technology Board also found water distribution systems difficult to secure and recognized that, while such systems may affect a smaller population, mass exposure is not needed if the terrorists' goal is fear and anxiety. As a result, public reports of illnesses may be the earliest indicator of deliberate contamination to distribution systems, according to one vulnerability assessment contractor.

A State water security coordinator said that neither EPA nor the different methodologies adequately emphasized distribution system threats as the most susceptible components of water systems to include in vulnerability assessments. While the RAM-W methodology acknowledges the susceptibility of threats to distribution systems, the methodology only mentions distribution systems as one of the many critical assets utility managers should seek to protect. Sandia's RAM-W program stems from a vulnerability assessment methodology initially developed to support the national nuclear security mission. The methodology has since been modified to evaluate the vulnerability to terrorist attack of government buildings, Air Force bases, nuclear power plants, nuclear processing facilities, prisons, and Federal dams. The State water security coordinator further said that RAM-W, as an artifact of nuclear- and dam-based methodologies, may be inappropriate for water utilities given their multiple facility size, unique and often elaborate distribution systems, and interconnections with other sectors.

Suggestions

EPA has plans to sample the vulnerability assessments to ensure compliance with Bioterrorism Act requirements. Based on our observations, we offer the following suggestions:

(1) EPA should consider including in its review a qualitative analysis of vulnerability assessments submitted by large utilities to determine whether they adequately address the threats envisioned by the Bioterrorism Act. Specifically, EPA's analysis should address whether the large utilities:

a. identified and prioritized specific threats – particularly terrorist scenarios; and

b. assessed the full breadth of a water system's infrastructure – particularly its distribution system.

(2) If EPA's analysis confirms our observations, EPA should focus on amending its guidance to address the shortcomings identified in this memorandum.

Agency Comments and Office of Inspector General Evaluation

In response to the concerns raised in our draft report, EPA analyzed a sample of the large water utility vulnerability assessments to determine if they specifically identified and addressed terrorist scenarios and distribution systems. EPA stated that any lessons learned from this analysis would be incorporated into guidance and training for medium and small water systems. Given that vulnerability assessments serve as the foundation for emergency response plans and future security enhancements, the OIG suggests that EPA monitor all water system submissions to ensure that vulnerability assessments identify and prioritize specific threats – particularly terrorist scenarios; and assess the full breadth of a water system's infrastructure – particularly its distribution systems.

The full Agency response is provided in Appendix A.

\- - -

If you or your staff have any questions regarding this report, please call me at (202) 566-0831.

Appendix A

Agency Response

UNITED STATES ENVIRONMENTAL PROTECTION AGENCY
WASHINGTON, D.C. 20460

JUN 16 2003

OFFICE OF
WATER

MEMORANDUM

SUBJECT: Response to OIG Concerns Regarding the Quality of Vulnerability Assessments Related to the Security of the Nation's Water Supply
DRAFT: Report No. 2003-M-000013

FROM: G. Tracy Mehan, III /s/
Assistant Administrator

TO: Jeffrey K. Harris
Director for Program Evaluation, Cross-Media Issues
Office of Inspector General

I am responding to the issues and concerns presented in your May 16, 2003, memorandum/report to me on the evaluation of the Environmental Protection Agency's (EPA) activities to enhance the security of the Nation's water supply. You specifically emphasized that EPA should promptly analyze the vulnerability assessments submitted by large utilities, as required by the Public Health Security and Bioterrorism Preparedness and Response Act of 2002 ("Bioterrorism Act"), to determine whether the assessments adequately and comprehensively address terrorist threats. In this analysis, you suggested that EPA consider a qualitative review of whether the large utilities:

a. identified and prioritized specific threats – particularly terrorist scenarios; and

b. assessed the full breadth of a water system's infrastructure – particularly its distribution system.

I understand that you made these suggestions because your preliminary research indicates there could be omissions in these areas. If that is the case, you propose that the Agency issue amended guidance to drinking water systems on conducting vulnerability assessments.

I believe that you and your staff would benefit greatly from a comprehensive briefing of the Water Protection Task Force's efforts over the past 20+ months. Your report focuses on the

Appendix A

reactions of numerous stakeholders to EPA's activities, so I encourage you and your staff to learn of them first hand from the Task Force's staff.

Development of Baseline Threat and Vulnerability Assessment Guidance

Since your memorandum/report cites interviews and discussions with many representatives in the drinking water community, I want to reiterate EPA's approach in developing the guidance to water utilities on assessing vulnerabilities to terrorist attacks and other intentional acts. First, the water industry, the federal public health, military, agricultural and food sectors, as well as the intelligence and law enforcement community were closely involved in identifying and defining risks to public health in relation to such attacks/acts. This was a critical step in both assisting utilities in developing their baseline threats for vulnerability assessments and in determining vulnerabilities in a distribution system relative to other infrastructure components of a water system. We also had a comprehensive process for developing tools and guidance on available baseline threat information that culminated in a meeting of national stakeholders in the Summer of 2002. This meeting was conducted to solicit feedback on issues relevant to conducting vulnerability assessments as well to reviewing and commenting on baseline threat concerns. In attendance were water industry officials from the Association of State Drinking Water Administrators (ASDWA), which represents the State primacy agencies, the American Metropolitan Water Agency (AMWA), which represents large water systems, and the American Water Works Association (AWWA), which represents the drinking water utilities, and managers/staff of several large municipalities including the Metropolitan Water District of Southern California and the City of Newport News, VA. The FBI sent experts from its National Infrastructure Protection Center to speak and act on behalf of the federal law enforcement sector and other sectors were represented by participants from the FDA, CDC, USDA and the US Army.

The primary purpose of this stakeholder meeting was to discuss a draft version of the *Baseline Threat Information for Vulnerability Assessments of Community Water Systems* (Baseline Threat Document) that was distributed to participants beforehand. This document was drafted to provide utility managers and their staffs with information necessary for the appropriate identification and evaluation of vulnerabilities, threats, and kinds of attack that could place the operation of the water utility (including the distribution system components), staff, and customers in harms way. One chapter of this document, *Determining The Level of Threat,* focused heavily on consideration of the terrorist threat as well as the national resources that are available to utilities to obtain threat information, e.g., the water information sharing and analysis center (WaterISAC). (Although in its infancy, the WaterISAC will provide utilities secure, timely, useable information to support efforts to protect the Nation's water infrastructure.) According to my staff, discussion of this chapter was active and intense especially around the FBI's assertion that intelligence on terrorist attacks is much more up-to-date and utility-specific at the field office level. As a result, the prevailing position of the stakeholders was that the design basis threat selection should be left to individual utilities to account for the uniqueness of each water system while incorporating the threat information gained from local FBI field offices and other security experts. Thus, this chapter in the final guidance presents a general description of the full range of threats, the historical threat perspective of the intelligence community (including input from the AWWArf and Sandia National Laboratories), and the specific recommendation that utilities seek participation and insight from local levels of law enforcement as they conduct their vulnerability assessments.

More extensive information - - in the form of appendices to the Baseline Threat Document - - on contamination threats and vulnerabilities was made available to utility managers. Most large utilities took advantage of this information as they conducted their vulnerability assessments as are medium and small systems that are currently conducting their assessments. While these materials cannot leave the secured area in which they are stored and filed, you and your staff can read and review this document and appendices by contacting the Water Protection Task Force.

Scope of Vulnerability Assessments

EPA agrees with the experts you interviewed that contamination of the distribution system could result in serious public health episodes. In our negotiations and discussions with the stakeholder organizations to which EPA provided financial support for the development of methodologies and tools for conducting vulnerability assessments, Agency officials highlighted this important area to reinforce and augment the RAM-W methodology with respect to distribution systems. An EPA official attended the train-the-trainer workshop and pointed out to the trainees that they would need to go beyond the focus of this methodology on distribution systems in order to consider and assess the vulnerabilities of the entire distribution system. At the same time, our intent for all the workshops we supported in 2002 was to concentrate on particular, high priority, areas of vulnerability in drinking water infrastructure and also give sufficient attention to all other areas as well. The Bioterrorism Act requires system evaluation as a whole and any emphasis on the distribution system without proper consideration or endorsement of the entire system could diminish the review of other vulnerable system components.

Also, I think it is important to recognize that security of the water sector, like all other sectors comprising homeland security, is a highly dynamic and evolving arena. Conducting vulnerability assessments should not be considered a one time endeavor but instead an iterative activity that water systems will have to review and update on a regular basis. EPA's approach to and support of current and future training on methodologies and tools for conducting vulnerability assessments of water systems will reflect "lessons learned" from 2002 and will incorporate state-of-the-art approaches developed in the interim.

Ongoing Assistance to Water Utilities

EPA's ongoing efforts in water infrastructure protection emphasize and support both research on contaminant monitoring approaches and technical assistance for water utilities and emergency response providers to act in response to contamination of water supplies. Workshops that will assist systems, serving between 50,000 and 100,000 people, in conducting their vulnerability assessments will be underway next month. Information and tools to address and strengthen action against identified vulnerabilities to attack and/or to disrupt water service entirely are continually being developed and implemented. For instance, the Agency is currently supporting the dissemination of a hydraulic model capable of predicting fate and transport of contaminants in distribution systems. This model is coupled with GIS tools to allow a system to identify locations that could be seriously affected by a "contamination event" and to develop appropriate proactive as well as response plans. A research strategy, developed jointly by the Office of Water, the Office of Research and Development and major stakeholders in the water community, will be published in the near future. This strategy contains an impressive mix of projects that cover a wide range of water security-related basic research as well as the development of technologies to detect, minimize, and protect against the introduction of harmful contaminants into water supplies.

Review of Vulnerability Assessments

I have already carried out one of the suggestions in your report. Staff (with top secret clearance) of the Water Protection Task Force has completed a qualitative review of a subset of the vulnerability assessments submitted by large drinking water systems. OW can brief you on the results of this review once you have been designated by the Administrator in accordance with the requirements of the Bioterrorism Act. As stated previously, my staff is ready to give you a full and detailed account of water security activities.

I appreciate the opportunity to respond to your draft report. Should you have any questions or need additional information, please contact Judy Hecht, the Office of Water's liaison to the IG's office, on 564-0475.

Appendix B

Distribution

Acting Administrator
Associate Administrator for Congressional and Intergovernmental Relations
Acting Associate Administrator, Office of Public Affairs
Assistant Administrator, Office of Water
Audit Followup Coordinator, Office of Water
Director, Office of Ground Water and Drinking Water
Chair, Water Protection Task Force
General Counsel
Director, Office of Homeland Security

Mrs. CAPPS. Mr. Chairman, securing the extensive network of our Nation's drinking water storage systems poses difficult challenges but the stakes are too high to shirk this responsibility. Tampering or destroying these systems would leave large population areas without water for consumption or fire-fighting purposes.

So I hope that precious time and money was not wasted by the EPA's failure to insure accurate vulnerabilities assessments keyed to post September 11 threats. Each day that passes without insuring that the necessary security enhancements are being undertaken is another day that our Nation's water supplies remain vulnerable.

So once again, this is a very timely hearing. I thank you for holding it and look forward to the testimony of our witnesses.

I yield back.

Mr. GILLMOR. The gentlewoman yields back. Does the gentleman from Nebraska have an opening statement?

Mr. TERRY. No.

Mr. GILLMOR. The gentleman yields.

The gentleman from New Jersey.

Mr. PALLONE. Thank you, Mr. Chairman.

I appreciate that you are giving us a chance in this subcommittee to conduct an oversight hearing on water security, which is one of the most pressing Homeland Security issues facing our country.

While it is certainly good that we are discussing water security, I want to express my dismay that this subcommittee, in fact, the entire House, has not held a single hearing to discuss security at the numerous chemical plants located in New Jersey and across the Nation.

As many of my colleagues may be aware, I have introduced H.R. 1861, the Chemical Security Act. This bill would, among other things, require that EPA promulgate regulations directing the owners of high priority chemical security plants to conduct vulnerability assessments and create a prevention, preparedness, and response plan much like the water programs that we will be discussing today.

But I can't emphasize enough the danger posed by chemical plants across the country. According to the EPA, there are 123 facilities across the country where release of chemicals could threaten more than 1 million people. There are more than 750 facilities where such a release could threaten upwards of 100,000 people—and these are frightening numbers. Despite this obvious threat, this subcommittee has done nothing on the issue. Last October, several of my colleagues joined me in then writing to then Chairman Tauzin asking that the committee address chemical security, but we have seen nothing since.

Through some combination of congressional and executive action, we have dealt port security, airline security and even nuclear security. If we are serious about securing our homeland and protecting our citizens, we need to address chemical security immediately.

Today's hearing—if I can say, Mr. Chairman, I look forward to the hearing and to hear what Mr. Grumbles and Mr. Stephenson have to stay. But, as we mentioned, this panel is clearly incomplete.

We are not going to hear from the EPA Inspector General, who issued two reports last September that were critical of the EPA's water security efforts. We are also not going to hear from the Department of Homeland Security, whom I understand simply refused to cooperate with the majority's request.

Yet this is a Congressional subcommittee. We have a serious oversight responsibility and the power to demand serious responses from the administration. And it is time this subcommittee and the Congress stand up to this administration. They are simply not cooperating. The Bush administration doesn't cooperate, whether it is water security or the lack of action on chemical security.

It is just a sad commentary on the state of oversight in this Congress, in contrast to the days when a Democrat was in the White House. This Congress has done very little to oversee the actions of the Bush Administration.

Mr. GILLMOR. The gentleman yields back.

Does the gentleman from New Hampshire have an opening statement.

Mr. BASS. I do, Mr. Chairman, I do have an opening statement I would like to have placed in the record.

Mr. GILLMOR. Without objection.

Mr. BASS. I would like to spend a minute and a half to tell an interesting little story.

We all remember 9/11 and what happened and how we found ourselves wandering around Capitol Hill looking for a place to go. I wound up at about noon, I think, in the headquarters of the Capitol Police over next to The Monocle on the Senate side. I was a motley crowd of Senators and Congressman sitting around the table looking kind of dazed. And the chief of the Capitol Police gave us a briefing, which was somewhat sketchy, about what he thought was going on. And Senator Byrd from West Virginia was there. And they had placed around the table all these pitchers of water and so forth for us to drink.

Senator Byrd said, what reason do you have to believe that the water supply in the District of Columbia hasn't been poisoned yesterday, and that we are—can't drink the water? We saw eight or 10 hands surreptitiously move across the table and push the pitchers to the other side of the table.

It only points out—that for me, at least, I realized just how significant protection of water resources can be in times of crisis. And this is a very timely hearing. Given the fact that only 15 percent of our water utilities control 75 percent of the population of this country, this is localized high priority for security. I think there are questions about whether or not it is practical or possible to really affect large segments of the population with any kind of ease, if you will. But nonetheless, I think this is an important subject and I commend the chairman for having this hearing and I yield back.

[The prepared statement of Hon. Charles F. Bass follows:]

PREPARED STATEMENT OF HON. CHARLES F. BASS, A REPRESENTATIVE IN CONGRESS FROM THE STATE OF NEW HAMPSHIRE

Thank you Mr. Chairman. It is important to note that the safety of our water supply is not a novel concern for our government. Since 1941, the federal government has recognized the possible threat to our national water supply—both by natural disasters and man-made threat. The fear of terrorist attacks on U.S. water utilities existed prior to 9/11 when EPA was identified as the leading federal agency to oversee security issues of our water infrastructure. Since the attack on American soil, the need to identify and address vulnerability has only become more imperative.

This hearing comes at a critical time due to the various legislative efforts in both Houses to address the recommendations of the 9/11 Commission. Since the creation of the Department of Homeland Security, EPA has continually retained their jurisdiction over water supply safety. It is critical for us to reassess how well this has worked with identifying potential threats, identifying and addressing points of vulnerability, and creating emergency plans. It is also important for us to include water supply safety in any decision that is related to coordination of intelligence and restructuring our agencies involved in homeland security.

Finally, this hearing is critical in discussing some of the types of potential threats that may exist to our water supply. Only 15% of our water utilities supply 75% of our population—and it is important for the public to understand the dangers that these utilities face. Some experts have argued that due to dilution and lengthy time for water to reach the home—the threat to public health is small. However, a threat to actually destroying the infrastructure is much greater. By including the public in these types of discussions, it helps elevate any unnecessary fears that may exist.

I would like to thank our witnesses for being here today and look forward to hearing their testimony. Thank you.

Mr. GILLMOR. The gentleman yields back. Does the other gentleman from Michigan have an opening statement?

Mr. STUPAK. Yes, I do, Mr. Chairman, and thank you for holding this hearing.

I want to thank Mr. Grumbles from the EPA and Mr. Stephenson from the Government Accountability Office for being here today as we discuss the safety and the security of our Nation's drinking water.

While I appreciate the chance to hear from our two witnesses today. I would have appreciated the chance to hear from anyone from the EPA's office of Inspector General as well considering the office's unique insight into this matter.

I would also like to hear testimony from the Department of Homeland Security, which plays a role in this issue, but has failed to respond to requests to testify before this subcommittee. The state of our Nation's drinking water supply is not something to be taken lightly. Clearly an attack on our Nation's water supply could have devastating consequences. That makes this hearing particularly important and the absence of witnesses from today's Homeland Security more troubling.

The passage of the Bioterrorism Act—following September 11, utilities serving a population of 3003 people were required to perform an submit an assessment of the water system's vulnerability to terrorist attacks or other attacks which might disrupt our drinking water supply. It is the EPA's responsibilities as the lead Federal agency charged with coordinating critical water infrastructure protection activities—to see that water utilities meet the requirements set forth by Congress in the Bioterrorism Act, as well as to provide them with the necessary tools to do so.

However, the EPA office of Inspector General has issued several reports in the last few years on this very subject critiquing the status of security at these very facilities.

In fact, the report released by the Inspector General last September found serious vulnerabilities in the EPA's actions to prevent or combat contamination of our water supply as a result of terrorism.

Specifically, EPA failed to provide adequate information about terrorist threats, but did provide guidance to water utilities on how to protect themselves from vandals. In short, the EPA was directing the local water utilities to be on the lookout for juvenile delinquents, not al Qaeda or terrorists. As a result, water utilities were having problems identifying and prioritizing threats to our water supply.

Given that vulnerability assessments serve as the foundation for emergency response plans and for future security enhancements, the Inspector General suggested that the EPA monitor all water system submissions to insure that vulnerability assessments identify and prioritize their terrorist threats.

In other words, EPA needs to make this a priority. We can't wait for an incident to happen. We need to take preventive action now. I hope the EPA will address the findings of the Inspector General today because these vulnerabilities are unacceptable.

With that, Mr. Chairman, I yield back the balance of my time.

[Additional statements submitted for the record follows:]

PREPARED STATEMENT OF HON. JOE BARTON, CHAIRMAN, COMMITTEE ON ENERGY AND COMMERCE

Thank you, Chairman Gillmor, for holding this oversight hearing today on a very critical issue affecting the health and security of our nation. Utilities across the country have long recognized that drinking water may be vulnerable to terrorism of various types, including physical disruption, bioterrorism, chemical contamination, and cyber attacks. I am proud to say that this Committee was the first to act on this issue after the attacks on September 11, with the passage of title IV of the Bioterrorism Act of 2002. Title IV amended the Safe Drinking Water Act to require each community water system serving more than the 3,300 individuals to conduct a "vulnerability assessment" of its susceptibility to a terrorist attack or other intentional act intended to substantially disrupt the ability of the system to provide safe drinking water.

It is my understanding that the timelines for all of these systems to have complied and submit their assessments have now passed. I look forward to getting an update on this process. While recognizing our drinking water systems' vulnerabilities is an important accomplishment, we also need to determine what steps are necessary in adopting appropriate security measures that address vulnerabilities and mitigate the consequences of any attack.

I thank the witnesses for their cooperation in attending and I look forward to hearing their testimony.

I yield back, Mr. Chairman.

PREPARED STATEMENT OF HON. JOHN D. DINGELL, A REPRESENTATIVE IN CONGRESS FROM THE STATE OF MICHIGAN

Mr. Chairman, I welcome this oversight hearing to determine the effectiveness of the Administration's implementation of the Safe Drinking Water Act Amendments that were enacted as part of the Public Health Security and Bioterrorism Preparedness and Response Act of 2002.

These provisions required drinking water utilities to assess the vulnerabilities of their distribution systems and water supplies to the potential threat of terrorist attacks. Water utilities were required to submit these assessments to the Environmental Protection Agency (EPA) so the government could ensure that they were properly conducted and that the drinking water utilities were taking the necessary actions to safeguard the public and protect drinking water supplies from potential terrorist threats.

While I look forward to Mr. Grumbles's testimony today, I am very disappointed that the Department of Homeland Security chose to ignore the Subcommittee's request to provide a witness. The hearing will also lack testimony from the EPA Inspector General's Office. This omission is particularly disappointing because the EPA Inspector General has issued four separate evaluation reports on EPA's performance and the assessments conducted by the water utilities.

The Inspector General's findings are extremely disturbing, and are worthy of this Subcommittee's careful review. For example, on September 23, 2003, the EPA Inspector General reported:

> "The Bioterrorism Act requires community water systems to prepare for and assess vulnerability to terrorist and other intentional acts. However, based on our interviews, we believe that vulnerability assessments submitted may emphasize traditional, less consequential, and less costly threats, such as vandalism or disgruntled employees. Therefore, vulnerability assessments may not necessarily address terrorist scenarios or the events of 9/11 that motivated passage of the Bioterrorism Act."

The Inspector General evaluation report dated September 11, 2003, stated:

> "EPA's Strategic Plan lacks fundamental components, such as measurable performance results and information and analysis, to ensure the greatest practicable reductions in risks to the critical water sector infrastructure."

If the vulnerability assessments are not addressing terrorist scenarios, and if EPA cannot demonstrate the risk reduction and security enhancements that have been achieved by water utilities, then the public interest is not being served.

I also note that while Congress has provided the Administration with the tools to assure and enhance security for water utilities, airlines, ports, and nuclear facilities, nothing has been done for chemical plants—one of our most vulnerable infrastructures that in the event of a terrorist attack could result in catastrophic loss of life. I urge the Committee to give this matter its full attention.

Mr. GILLMOR. The gentleman yields back. The Chair will recognize himself for some questions, after our witnesses testify.

Mr. Grumbles.

STATEMENT OF BENJAMIN H. GRUMBLES, ACTING ASSISTANT ADMINISTRATOR FOR WATER, U.S. ENVIRONMENTAL PROTECTION AGENCY

Mr. GRUMBLES. Thank you, Mr. Chairman, and Congresswoman Solis, and members of the subcommittee. I am Ben Grumbles with the office of water EPA. I am here to talk a little bit about the progress we have made in the Bioterrorism Act of 2002, the partnerships that have allowed progress to be made and also our priorities and the challenges we face—and Mr. Chairman, I would be remiss if I did not acknowledge, since I know personally firsthand the role of this subcommittee and this committee in crafting the bipartisan legislation, the Bioterrorism Act of 2002 and moving forward with it.

And I can say that the administration is very proud of the legislation and the success we have had to date in implementing it.

I can also say—that while I will talk about some of the statistics in terms of the vulnerability assessments and the emergency response plans—that I agree full well with the spirit and the tone of some of the statements that assessments and plans by themselves do not make systems safer.

However, I can also say with confidence that—on the water security front—we are smarter and safer as a country than we were 3 years ago.

And a lot of that is due to the legislation and also, quite frankly, the aggressive efforts of the EPA and the administration to implement the legislation and to do things outside of the legislation.

And perhaps most importantly, it has been the commitment and the efforts of the utilities, the local officials, the State drinking water agencies and others, some of the folks are in the room behind me, to really move forward in terms of the Bioterrorism Act.

Well, what I would like to do is to focus on a couple of the aspects, Title IV of the Bioterrorism Act that you all were instrumental in drafting and enacting and overseeing.

First of all, we have some excellent numbers to report on the vulnerability assessments. The first number is 100 percent. That is the number that reflects the compliance rate of the large drinking water systems throughout the country. They have all submitted their vulnerability assessments to EPA.

The other number I would like to mention is 98 percent. That is the number of medium-sized community water systems that have submitted their vulnerability assessments. 89 percent. That is the number of emergency response plan certifications that have been submitted by the medium-sized communities.

The last number is 88 percent, and that is the number of small systems—those between 3,300 population and 50,000 in population—that have submitted their vulnerability assessments.

And, Mr. Chairman, what these numbers mean is that the country is listening, the utilities throughout the country using our guidance and following the law have submitted their vulnerability assessments. There is a much greater awareness, and they have also largely submitted all of their certifications with respect to the emergency response plans. I would also like to say that we have partnered with the domestic preparedness office of the Department of Homeland Security to offer workshops to train drinking water utilities on emergency response planning.

What are some of the priorities and activities that we are focusing on? One of them, which is critically important, is to move beyond just the identifying risks, doing vulnerability assessments and preparing emergency response plans. A high priority of the agency is to provide the tools, the training, the technical assistance to actually implement those plans. That means taking measures of prevention, hardening facilities, taking various steps to insure that the drinking water systems throughout the country are truly safer and more secure.

I think it is critically important that water utilities stay up to date on the threat information. The agency has been providing threat information, our baseline threats documents or guidance to utilities on preparing their vulnerability assessments are actions that we take pride in and recognize can help our partners do the important job they need to do.

I just want to mention a couple other items, Mr. Chairman, that are significantly important. One of those is that the Agency is working on implementation of Presidential Directive 9. It is the homeland security Presidential Directive number 9 that was issued in January 2004.

That is a comprehensive and ambitious directive to us to improve and increase the monitoring and surveillance of the Nation's drinking water systems. Monitoring is critically important and we are putting a high priority on working with our partners and our other members of the Federal family, certainly the Department of Homeland Security to follow through on the President's directive.

The other thing I wanted to mention is an excellent example of the partnerships that are critically important to the success of water security, and that is we are working with the American Society of Civil Engineers to develop physical security guidelines that utilities should consider in designing, managing and operating their systems.

Mr. Chairman, there are over 2 million miles of pipe in the country with respect to drinking water facilities. What that tells all of us is that one of the priority areas in implementing your legislation, our legislation, is to focus in on the distribution systems, that is a primary focus, and we will certainly continue to do that through our research plan, through our actions and through coordinating our responses under the Presidential directives.

The last thing I want to mention, Mr. Chairman, are of the challenges and opportunities. Several of your colleagues have mentioned some of the key issues, and I would like to reiterate them.

One of the challenges is to recognize that the vulnerability assessments should be living documents. The visions, the great legislation of the 2002 Bioterrorism Act essentially left it that those were one-time assessments. I think what we have learned in our coordinations with other partners with GAO, with the Inspector General, is that there would be great value if those documents were living documents and would be revisited and revised and updated and adapted, modernized. So that is a very important thing to keep in mind.

The other one, the final one, Mr. Chairman, is the vulnerability assessments themselves. We think it is very important for you and your colleagues to keep in mind the delicate balance of insuring the security of those assessments, certainly as it relates to site specific information. Again, what the Inspector General told us and what we very much appreciate hearing—and what GAO and others tell us is that—there can be value to aggregate data based in general on the vulnerability assessments that can help shed information and light on our research plans, our priorities. That is another thing for the committee to keep in mind.

Mr. Chairman, I thank you and your colleagues for your patience, and we look forward to answering any of your questions.

[The prepared statement of Benjamin H. Grumbles follows:]

PREPARED STATEMENT OF BENJAMIN H. GRUMBLES, ACTING ASSISTANT ADMINISTRATOR FOR WATER, U.S. ENVIRONMENTAL PROTECTION AGENCY

INTRODUCTION

Good afternoon Chairman Gillmor and Members of the Committee. I am Benjamin H. Grumbles, Acting Assistant Administrator for Water at the United States Environmental Protection Agency. I welcome this opportunity to speak to you today about our progress to date in water security, our vision for the future, and the challenges we face in enhancing the security of the Nation's water infrastructure.

Promoting the security of the Nation's water infrastructure is one of the most significant undertakings and responsibilities of the Agency in a post-September 11 world. An attack, or even a credible threat of an attack, on water infrastructure could seriously jeopardize the public health and economic vitality of a community. As you know, drinking water and wastewater utilities can be vulnerable to a variety of attacks, including, for example, physical destruction of critical water system components, release of hazardous chemicals, intrusion into cyber systems, and intentional contamination of drinking water.

Over the past three years, EPA has worked diligently to support the water sector in improving water security and the sector has taken their charge seriously. Through Congressional authorization under the *Public Health Security and Bioterrorism Preparedness and Response Act of 2002* (the Bioterrorism Act), and through Presidential mandates under Homeland Security Presidential Directives 7, 9 and 10, EPA has been entrusted with important responsibilities for coordinating the protection of the water sector.

We have good news to report on our progress to date. However, much work remains to be done. Understanding one's vulnerability is only the first step in what is a multi-step process to improving security. Many water systems that have completed their vulnerability assessments are now saying, "we have identified our weaknesses, now what do we do?" The next steps involve adopting security measures that both address vulnerabilities and mitigate the consequences of an attack.

EPA's water security work has focused on helping utilities assess their vulnerabilities and creating a baseline of security-related information. Existing and future efforts include providing tools and assistance that drinking water and wastewater systems need to address vulnerabilities by identifying up-to-date security enhancements, sharing information on threats and contaminants, and training on emergency response.

Our goal is to provide the water sector and related emergency response, law enforcement, and public health officials with the tools, training, and information they need to prevent, prepare, and respond to terrorist threats. EPA also needs to continue to provide programs that forge critical links between the water sector and those who support or could support the sector in detecting and responding to threats and incidents, such as local law enforcement and public health departments.

Indeed partnerships are absolutely a key factor in our success. The water sector includes approximately 54,000 community drinking water systems and 16,000 publicly owned wastewater treatment works nationwide. Reaching the entire water sector requires strong partnerships among EPA, state water and homeland security officials, and technical assistance providers. Our work also demands extensive coordination and communication among federal agencies including the Department of Homeland Security, the Department of Health and Human Services, the Department of Defense and the intelligence community, among others.

As a result of the partnerships we have developed and EPA's long-standing relationship with the water sector, we have fulfilled the requirements of the Bioterrorism Act of 2002 and made headway on several other fronts, as well.

IMPLEMENTATION OF TITLE IV B DRINKING WATER SECURITY AND SAFETY

Required Vulnerability Assessments and Emergency Response Plans

Under the Act, each community water system (CWS) providing drinking water to more than 3,300 persons must conduct a vulnerability assessment, certify its completion, and submit a copy of the assessment to EPA according to a specified schedule. In addition, each system must prepare or revise an emergency response plan that incorporates the findings of the vulnerability assessments and certify to EPA within six months of completing a vulnerability assessment that the system has completed such a plan.

Using FY 2002 supplemental appropriation funds, EPA provided grants to support the development of vulnerability assessments and emergency response plans. EPA issued $51 million in direct grants to 399 of the largest community water utilities that serve populations greater than 100,000 people. Working with training organizations and State drinking water administrators, EPA provided $20 million in grants to provide technical assistance to small and medium community water systems.

EPA has received all of the vulnerability assessments and emergency response plan certifications from the Nation's largest community water systems. To date, we have received vulnerability assessments from 98% of the medium-sized community water systems that were due December 31, 2003, and 89% of their emergency response plan certifications. The smallest community water systems covered by the Act were required to submit their vulnerability assessments to us by June 30, 2004. We have received over 7,000 vulnerability assessments from this group, amounting to an 88% submission rate. What these numbers mean is that water systems serving collectively over 230 million people have completed vulnerability assessments: a remarkable achievement in so short a time. Despite this success, EPA continues to work to ensure that we receive all vulnerability assessments and emergency response plan certifications so that all of the Nation's community water systems serving more than 3,300 people reach the same critical milestone.

Of course, most of the credit should go to those who actually prepared the vulnerability assessments and emergency response plans: the water systems themselves. Without their commitment to enhancing security for their consumers, we would not have seen such a high response rate.

Information on Baseline Threats and Protection Protocols

The Bioterrorism Act also required EPA to develop and provide baseline threat information to community water systems in order to aid them in performing vulnerability assessments. EPA developed the *Baseline Threat Information for Vulnerability Assessments of Community Water Systems* (Baseline Threat Document) in consultation with many stakeholders, including other federal agencies, state and local governments, water industry associations, and technical experts. The Baseline Threat Document provides utilities with information to (1) undertake risk-based vulnerability assessments of their assets, (2) analyze potential threats, and (3) consider the consequences of a variety of modes of attack. The document, whose distribution is limited largely to community water systems, lists vulnerability assessment tools and other information resources to help water systems learn more about the potential threats in their areas.

To further assist community water systems in completing their vulnerability assessments and emergency response plans, in January 2003, EPA released a document titled, *Instructions to Assist Community Water Systems in Complying with the Public Health Security and Bioterrorism Preparedness and Response Act of 2002.* An addendum to the instructions was released in October 2003. The instructions outline the steps that water utilities should take to transmit their vulnerability assessments and certifications to EPA. The instructions and a supporting fact sheet also outline the six key elements and all components of the system, as specified in the Act that must be considered in the vulnerability assessment.

Besides the commitment of the utilities and Congressional support for funding, we attribute the success in meeting the requirements of the Act to several factors. First, to aid the development of vulnerability assessments and emergency response plans, EPA supported the creation of analytical tools, training, and technical assistance for the range of sizes of drinking water systems. Vulnerability assessment tools include the Risk Assessment Methodology for Water, which has since been adapted for small and medium drinking water utilities; the CD-ROM software Vulnerability Self-Assessment Tool for drinking water and wastewater systems; and Security and Emergency Management System for small drinking water systems.

Second, working with our many partners, EPA-sponsored training and workshops in 2002 and 2003 which reached several thousand community drinking water and wastewater utility officials, training providers, and utility contractors. These efforts have trained drinking water and waste water systems that serve most of the U.S. population.

To aid the development of emergency response plans, as required by the Act, EPA developed guidance outlining the elements of a sound plan followed by a toolbox entitled the *Response Protocol Toolbox: Planning and Responding to Contamination Threats to Drinking Water Systems,* which is designed to help utilities prepare for and respond to intentional contamination threats and incidents.

Over the past year, EPA has partnered with DHS's Office of Domestic Preparedness to offer a series of workshops to train drinking water utilities on emergency response planning. A series of two-day workshops feature a tabletop exercise of an intentional contamination event in a public water supply. The goal of the exercise is to bring representatives of the key response agencies (e.g., FBI, local and state police, emergency responders, state regulatory agencies, state and local health departments) together to apply the guidance provided during the first day of training.

While EPA has worked to ensure that community water systems fulfill their obligations under the Bioterrorism Act, the Agency has not ignored wastewater systems or small community drinking water systems (serving 3,300 and fewer), which are not subject to specific provisions of the Bioterrorism Act requiring the completion of vulnerability assessments and emergency response plans. EPA also has provided guidance and training to these utilities on how to conduct vulnerability assessments, prepare emergency response plans, and address threats from terrorist attacks.

Research

The Act also places a premium on ensuring that research is carried out to support security efforts. Section 1434 of the Act stipulates that EPA shall work collaboratively to review methods to prevent, detect, and respond to the intentional contamination of water systems, including a review of equipment, early warning notification systems, awareness programs, distribution systems, treatment technologies and biomedical research. Section 1435 requires the review of methods by which the water system and all its parts could be intentionally disrupted or rendered ineffective or unsafe, including methods to interrupt the physical infrastructure, the computer infrastructure, and the treatment process.

To meet EPA's mandate under these sections, the Office of Water partnered with the newly established National Homeland Security Research Center in EPA's Office of Research and Development to draft the *Water Security Research and Technical Support Action Plan*. The Action Plan, released in March 2004, addresses each of the research requirements under the Bioterrorism Act. It describes the research and technologies needed to better address drinking water supply, water treatment, finished water storage, and drinking water distribution system vulnerabilities. It also addresses water security research needs for wastewater treatment and collection infrastructure, which includes sanitary and storm sewers or combined sanitary-storm sewer systems, wastewater treatment, and treated wastewater discharges. EPA is implementing activities described in the plan, which was vetted with water stakeholders and reviewed by the National Academy of Science.

FULFILLING OUR GOAL: ACTIVITIES, PLANS AND CHALLENGES

As I mentioned earlier, our goal is to provide the water sector the tools, training, and information they need to comprehensively address water security. With utilities and our other partners, we are aiming to minimize the opportunity for terrorist attack on drinking water or wastewater systems by identifying and reducing potential risks and to maximize our ability to detect and respond to terrorist attacks. Let me give you some examples of the activities we have underway and challenges we face to support this goal.

Identifying Risk

In addition to undertaking vulnerability assessments, it is vital that water utilities stay up-to-date on threat information in order to fully understand their potential risk. Funded in large part by EPA, the Water Information Sharing and Analysis Center, known as the WaterISAC, became operational in December 2002. It was developed to provide drinking water and wastewater systems with a highly secure Web-based environment for early warning of potential physical, contamination, and cyber threats and for information about security. The 311 utilities that currently subscribe to the WaterISAC provide drinking water to 60 percent of the U.S. population. Forty-five State drinking water primacy agencies are members of the WaterISAC, which provides a mechanism to reach the majority of small and medium drinking water systems. Key EPA staff also have access.

Efforts are underway to expand membership in the WaterISAC and to develop the ancillary Water Security Channel (WaterSC) that will allow the WaterISAC to send e-mail alerts on security issues and share basic security information directly with a much larger group of drinking water and wastewater systems.

Recently, the Department of Homeland Security announced plans to expand its secure, computer-based counter-terrorism network to the critical infrastructures, working first with the water and electricity sectors. The National Homeland Security Information Network (HSIN) reaches state homeland security offices, emergency operations centers around the country, and has a significant law enforcement communications component. EPA is working with the appropriate organizations to determine how the WaterISAC and HSIN can best serve water sector utilities.

In addition, EPA works with the Department of Homeland Security and the broader intelligence community to improve threat information relevant to water utilities. This involves training intelligence officers on the vulnerabilities of water utilities and providing secure mechanisms, such as the WaterISAC, to communicate sensitive information to the utilities.

Reducing Risk

Early warning mechanisms can significantly reduce the risk of public health impacts and community service disruptions. Issued in January 2004, Homeland Security Presidential Directive (HSPD 9) outlines EPA's responsibilities to develop a robust, comprehensive surveillance and monitoring program to provide early warning in the event of a terrorist attack using biological, chemical, or radiological contaminants. HSPD 9 also directs EPA to develop a nationwide laboratory network to support the routine monitoring and response requirements of the surveillance program.

EPA worked closely with water utilities, state officials and other federal agencies, for example the Department of Health and Human Services, the Department of Homeland Security and the Department of Defense, to formulate the conceptual framework for building such a surveillance and laboratory capability. Specific activities supporting this analysis included: 1) development of a standardized field screening and sampling kit; 2) identification of the highest priority contaminant threats and the most vulnerable infrastructure points through an inter-agency workgroup, 3) evaluation of new and emerging detection technologies; and 4) collaboration with the Centers for Disease Control and Prevention (CDC) to develop an alliance of drinking water laboratories with CDC's Laboratory Response Network.

In recognition that a robust detection program is only one part of an effective security strategy, EPA developed a variety of policies, procedures, physical enhancements, and best practices that assist water utilities in preventing attacks and protecting critical infrastructure components. For example, EPA's Security Product Guides provide information on a variety of products available to enhance physical security (including monitoring equipment) and electronic or cyber security. Several products will assist utilities in preventing or delaying potential adversaries as well as detecting incidents. In addition, EPA has worked with the American Society of Civil Engineers to develop physical security guidelines that utilities should consider in designing, managing, and operating their systems.

Implementing security enhancements can prove to be a challenge for many water-sector utilities who also face competing demands for replacement of aging infrastructure and making process improvements to meet public health requirements. EPA and water-sector stakeholders need to continue educating elected officials, water boards, rate-setting entities, and consumers about the importance and need for security enhancements at drinking water and wastewater utilities and the multiple benefits that can be derived from these enhancements. EPA has provided guidance on how the Drinking Water State Revolving Fund and the Clean Water State Revolving Fund may be used to lend financial support for such improvements.

Preparing to Respond

Due to the dispersed nature of water utilities B the Nation's drinking water utilities have about 2 million miles of pipe B it is a great challenge to protect against determined aggressors. Consequently, it is critically important that water utilities be prepared to respond effectively at any time. Building on workshops already given in FY 2003 and FY2004, EPA will continue to stress the importance of emergency response planning, drills and exercises for water utilities and associated emergency response, law enforcement and public health officials.

Several Homeland Security Presidential Directives (HSPDs) issued within the year also relate to emergency response. For example, HSPD 8 (December, 2003) establishes policies to strengthen the Nation's preparedness to prevent and respond to threatened or actual domestic terrorist attacks, major disasters, and other emergencies by establishing mechanisms for improved delivery of federal preparedness assistance to state and local governments. Also, HSPD 10: Biodefense for the 21st Century (April, 2004), which is currently a classified document, reaffirms EPA's responsibilities under HSPD 9 while adding a clear directive on the Agency's responsibilities in decontamination efforts. It provides direction to further strengthen the Biodefense Program through threat awareness, prevention and protection, surveillance and detection, and response and recovery.

CHALLENGES AND OPPORTUNITIES

While progress has been made toward securing drinking water and wastewater utilities, a number of challenges and opportunities remain, and EPA is taking steps to meet them both from national and local perspectives .

EPA was designated as the Sector Specific Agency responsible for infrastructure protection activities for the nation's drinking water and wastewater systems under HSPD 7, *entitled Critical Infrastructure Identification, Prioritization, and Protection* (December, 2003). As such, EPA is responsible for: 1) identifying, prioritizing, and coordinating infrastructure protection activities for the nation's drinking water and wastewater treatment systems; 2) working with federal departments and agencies, state and local governments, and the private sector to facilitate vulnerability assessments; 3) encouraging the development of risk management strategies to protect against and mitigate the effects of potential attacks on critical resources; and 4) developing mechanisms for information sharing and analysis. As I have explained, work is underway to fulfill many of these responsibilities.

To portray a comprehensive picture of security activities for the water sector, under HSPD 7, EPA is leading the development of a water sector specific plan as part of the DHS-led National Infrastructure Protection Plan production process.

In developing the plan, we identified some additional issues for ensuring that water utilities implement effective security programs. For example, updates of drinking water utilities' vulnerability assessments and emergency response plans, or the implementation of security enhancements identified by the vulnerability assessment, are not required. The water sector recognizes the need for both vulnerability assessments and emergency response plans to be living documents, revised periodically to ensure their applicability. Furthermore, sector representatives have expressed to the Agency the need for clear expectations of what constitutes effective security programs so that they can justify and obtain the resources needed to improve security.

To address this challenge, the Agency asked the National Drinking Water Advisory Council (NDWAC), a formal advisory committee to the Agency, to consider establishing a Water Security Working Group to (1) characterize effective voluntary utility security programs for drinking water and wastewater utilities, (2) consider ways to provide recognition and incentives that facilitate adoption of such programs, and (3) recommend mechanisms to measure the extent of implementation. The NDWAC agreed and the resultant Working Group is made up of sixteen members chosen on the basis of experience, geographic location, and their unique drinking water, wastewater, and/or security perspectives. During the first meeting of the workgroup, it was clear that the Working Group will consider the need for an iterative approach whereby utilities periodically revisit both vulnerability assessments and emergency response plans.

Another issue that we identified relates to EPA's ability to share the information contained in or derived from vulnerability assessments that are required by the Act to be submitted to the Agency by Community Water Systems. Currently, consistent with the protective provisions of the Bioterrorism Act, EPA must designate individuals before sharing assessment information with them. Clearly, it is extremely important to protect the site-specific vulnerability information contained in these vulnerability assessments and the Agency guards this information fiercely. Aggregated information on vulnerabilities of the sector, however, could be helpful in identifying priorities for security improvements and research. Both the Government Accountability Office and EPA's Inspector General have pointed out the need for this information to guide our efforts at the federal level.

CONCLUSION

EPA has developed a water security program that meets our critical responsibilities as expressed in Homeland Security Presidential Directive 7, which assigns to EPA a pivotal role in coordinating and facilitating the protection of the Nation's drinking water and wastewater systems. EPA has produced a broad array of tools and assistance that the water sector is using to assess its vulnerabilities and to develop emergency response plans. As a result of our efforts, drinking water systems collectively serving over 230 million people have submitted vulnerability assessments. We have worked effectively with our partners within the sector and also reached out to build new relationships with important partners beyond the sector to ensure that water and wastewater utilities receive the information and support they need to reduce risk and consequences of an attack.

Thank you for the opportunity to describe our accomplishments, new mandates and program needs, challenges, and vision for the future of water infrastructure security. Looking forward, we will continue to work closely with Congress, our water sector partners, federal agencies and various stakeholders to ensure that citizens across the country are confident in the security of their water and wastewater utilities. I will be happy to answer any questions you may have.

Mr. GILLMOR. Thank you very much, Mr. Grumbles.

Mr. Stephenson.

STATEMENT OF JOHN B. STEPHENSON, DIRECTOR, NATURAL RESOURCES AND ENVIRONMENT, GOVERNMENT ACCOUNTABILITY OFFICE

Mr. STEPHENSON. Thank you, Mr. Chairman, members of the subcommittee.

Drinking water utilities have long been recognized as potentially vulnerable to terrorist attacks of various types, including physical disruption, bioterrorism, chemical contamination and cyber attack. Terrorists could disrupt not only the availability of safe drinking water, but also the delivery of vital services that depend on these water supplies, such as fire suppression. Such concerns were greatly amplified by the September 11 attacks on the World Trade Center and the Pentagon—and then by the discovery of training manuals in Afghanistan, detailing how terrorist trainees could support attacks on drinking water systems.

Congress, as you know, has committed over $140 million in fiscal years 2002 through 2004 to help systems assess their vulnerability to terrorist threats and to develop response plans.

My testimony today is based on a report that we did last year on how best to use these funds. To develop this report, we examined the key security-related vulnerabilities affecting the Nation's drinking water systems; how Federal funds could best be used; and, specific activities that the Federal Government should support to improve drinking water security.

To address these issue, we assembled a panel of 43 nationally recognized experts and in selecting these experts we sought individuals who were widely recognized as possessing expertise on one or more key aspects of drinking water security. We also sought to achieve a balance in representation from key Federal agencies, key State and local agencies, industry and nonprofit organizations and water utilities of various sizes. Here is what our experts said.

Nearly 75 percent of the experts identified the distribution system as the most vulnerable of all system components with source water supplies, critical information or data systems and chemicals stored onsite as the next most important vulnerability. A typical drinking water system with a supply source water facility and distribution system—the distribution system was cited as the greatest vulnerability because it is easily accessible at so many points, such as a fire hydrant or a standpipe within a building.

In fact, the water is post treatment, meaning that a chemical, biological or radiological agent would be virtually undetectable until it was too late to prevent harm.

The experts also identified a lack of redundancy in biosystems and a lack of information on the most serious threats as overarching vulnerabilities.

In responding to our questions about how Federal funds could best be used, about 90 percent of our experts said that allocation decisions should be based on the vulnerabilities assessments prepared under Bioterrorism Act.

In addition, the experts gave the highest funding priority to utilities serving high density populations followed closely by utilities serving critical assets such as military bases or other sensitive government utilities.

When asked to identify the most effective mechanisms for distributing these Federal funds—over half the experts favored direct Federal grants—but many favored a Federal requirement for matching funds pass a grant condition.

Fewer experts recommended that using the drinking water State resolving fund, cautioning that it would not be as effective for making near-term security upgrades, and that it might dilute the fund's original purpose of infrastructure upgrade.

Finally, when we ask our experts to identify and set priorities for security-enhancing activity, most deserving of Federal support, their responses fell into three categories. The first was physical, and physical improvements including the development of real-time monitoring technologies, increasing laboratory capacity and physical hardening.

The second was education and training to be provided to both utility and nonutility personnel responsible for preventing, responding to and recovering from an attack.

And three, strengthened operational relationships, especially between water utilities and other agencies such as public health, enforcement agencies, and neighborhood utilities that may have a key role in emergency response.

In conclusion, we recommended that EPA consider the information in this report as it determines how best to allocate security-related Federal funds among drinking water utilities, which method should be used to distribute the funds and what specific activities should be supported. EPA agreed to do so. As it moves forward with the drinking water security program, we think it is doing so.

Mr. Chairman, that concludes the summary of my statement and I will take questions.

[The prepared statement of John B. Stephenson follows:]

PREPARED STATEMENT OF JOHN B. STEPHENSON, DIRECTOR, NATURAL RESOURCES AND ENVIRONMENT, U.S. GOVERNMENT ACCOUNTABILITY OFFICE

Mr. Chairman and Members of the Subcommittee: Drinking water utilities across the country have long been recognized as potentially vulnerable to terrorist attacks of various types, including physical disruption, bioterrorism, chemical contamination, and cyber attack. Damage or destruction by terrorists could disrupt not only the availability of safe drinking water, but also the delivery of vital services that depend on these water supplies, such as fire suppression. Such concerns were greatly amplified by the September 11, 2001, attacks on the World Trade Center and the Pentagon and then by the discovery of training manuals in Afghanistan detailing how terrorist trainees could support attacks on drinking water systems.

Congress has since committed significant federal funding to assist drinking water utilities—with over $140 million appropriated from fiscal year 2002 through fiscal year 2004—to help systems assess their vulnerabilities to terrorist threats and develop response plans. As significant as these funds are, drinking water utilities are asking the federal government to support efforts that go beyond the *planning* for upgrading drinking water security to the actual *implementation* of security upgrades. Consequently, at the request of the Senate Committee on Environment and Public Works, we examined (1) the key security-related vulnerabilities affecting the nation's drinking water systems; (2) the criteria that experts believe should be used to determine how federal funds are allocated among recipients to improve their security, and the methods that should be used to distribute these funds; and (3) specific activities that experts believe the federal government should support to improve drinking water security. My testimony is based on our October 2003 report entitled, *Drinking Water: Experts' Views on How Future Federal Funding Can Best Be Spent to Improve Security.*

To prepare our October 2003 report on these issues, we assembled a panel of nationally recognized experts. In selecting members for the expert panel, we sought individuals who were widely recognized as possessing expertise on one or more key aspects of drinking water security. We also sought to achieve balance in representation from key federal agencies, key state or local agencies, key industry and nonprofit organizations, and water utilities of varying sizes.

In summary:

- Our expert panel identified several key physical assets as the most seriously vulnerable to terrorist attacks. Nearly 75 percent of the experts (32 of 43) identified one or more components of the distribution system. In fact, more experts identified the distribution system as the single most important vulnerability (12 of 43) of all system components. The other physical assets most frequently cited were source water supplies, critical information systems, and chemicals that are stored on site for use in the treatment process. Importantly, the experts also identified overarching vulnerability issues that may involve multiple system components, or even an entire drinking water system. Chief among these issues were (1) a lack of redundancy in vital systems, which increases the likelihood that an attack could render a system inoperable; and (2) the difficulty many systems face due to a lack of information on the most serious threats to which they are exposed. In general, the panelists' observations were similar to those of major public and private organizations that have assessed the vulnerability of these systems to terrorist attacks, including the National Academy of Sciences, Sandia National Laboratories, and key industry associations.

- About 90 percent of the experts agreed "strongly" or "somewhat" that allocation decisions should be based on assessments of drinking water utilities' vulnerabilities, which the utilities are required to prepare by the Public Health Security and Bioterrorism Preparedness and Response Act of 2002. In addition, the experts favored funding priority for utilities serving high-density populations, with over 90 percent indicating that they deserve at least a "high" priority and over 50 percent indicating they deserve "highest" priority. Utilities serving critical assets (such as military bases and other sensitive government facilities, national icons, and key cultural or academic institutions) were also recommended as high-priority recipients. When asked to identify the most effective mechanisms for distributing these federal funds to recipients, over half the experts indicated that direct federal grants would be "very effective" in doing so. Many also favored including a requirement for matching funds as a grant condition. Fewer experts recommended using the Drinking Water State Revolving Fund (DWSRF) for this purpose, particularly to support upgrades that need to be implemented quickly.
- When asked to identify and set priorities for security-enhancing activities most deserving of federal support, the experts most frequently identified activities that fell into three broad categories:
  - *Physical and technological improvements*—needed for both physical alterations to improve the security of drinking water systems, and for the development of technologies to prevent, detect, or respond to an attack. The need to develop near real-time monitoring technologies, which would be particularly useful in quickly detecting contaminants in water that has already left the treatment plant for the consumer, had by far the strongest support.
  - *Education and training*—to be provided to both utility and nonutility personnel responsible for preventing, responding to, and recovering from an attack. These activities include, among other things, support for simulation exercises to provide responders with experience in carrying out utilities' emergency response plans; specialized training of utility personnel responsible for security; general training of utility personnel to augment security awareness among all staff; and multidisciplinary consulting teams to independently analyze utilities' security preparedness and recommend security-related improvements.
  - *Strengthened operational relationships*—especially between water utilities and other agencies (public health agencies, enforcement agencies, and neighboring utilities, among others) that may have key roles in an emergency response. This category also includes developing common protocols to engender a consistent approach among utilities in detecting and diagnosing threats, and the testing of local emergency response systems to ensure that participating agencies coordinate their actions effectively.

BACKGROUND

Drinking water systems vary by size and other factors, but as illustrated in figure 1, they most typically include a supply source, treatment facility, and distribution system. A water system's supply source may be a reservoir, aquifer, or well, or a combination of these sources. Some systems may also include a dam to help maintain a stable water level, and aqueducts and transmission pipelines to deliver the water to a distant treatment plant. The treatment process generally uses filtration, sedimentation, and other processes to remove impurities and harmful agents, and disinfection processes such as chlorination to eliminate biological contaminants. Chemicals used in these processes, most notably chlorine, are often stored on site at the treatment plant. Distribution systems comprise water towers, piping grids, pumps, and other components to deliver treated water from treatment systems to consumers. Particularly among larger utilities, distribution systems may contain thousands of miles of pipes and numerous access points.

Nationwide, there are more than 160,000 public water systems that individually serve from as few as 25 people to 1 million people or more. As figure 2 illustrates, nearly 133,000 of these water systems serve 500 or fewer people. Only 466 systems serve more than 100,000 people each, but these systems, located primarily in urban areas, account for early half of the total population served.

Until the 1990s, emergency planning at drinking water utilities generally focused on responding to natural disasters and, in some cases, domestic threats such as vandalism. In the 1990s, however, both government and industry officials broadened the process to account for terrorist threats. Among the most significant actions taken was the issuance in 1998 of Presidential Decision Directive 63 to protect the nation's critical infrastructure against criminal and terrorist attacks. The directive designated the Environmental Protection Agency (EPA) as the lead federal agency to address the water infrastructure and to work with both public and private organizations to develop emergency preparedness strategies. EPA, in turn, appointed the Association of Metropolitan Water Agencies to coordinate the water industry's role in emergency preparedness. During this time, this public-private partnership focused primarily on cyber security threats for the several hundred community water systems that each served over 100,000 persons. The partnership was broadened in 2001 to include both the drinking water and wastewater sectors, and focused on systems serving more than 3,300 people.

Efforts to better protect drinking water infrastructure were accelerated dramatically after the September 11 attacks. EPA and the drinking water industry launched efforts to share information on terrorist threats and response strategies. They also undertook initiatives to develop guidance and training programs to assist utilities in identifying their systems' vulnerabilities. As a major step in this regard, EPA supported the development, by American Water Works Association Research Foundation and Sandia National Laboratories, of a vulnerability assessment methodology for larger drinking water utilities. The push for vulnerability assessments was then augmented by the Public Health Security and Bioterrorism Preparedness and Response Act of 2002 (Bioterrorism Act). Among other things, the act required each community water system serving more than 3,300 individuals to conduct a detailed vulnerability assessment by specified dates in 2003 or 2004, depending on their size.

Since we issued our report in October, several Homeland Security Presidential Directives (HSPDs) were issued that denote new responsibilities for EPA and the water sector. HSPD 7 designates EPA as the water sector's agency specifically responsible for infrastructure protection activities, including developing a specific water sector plan for the National Infrastructure Protection Plan that the Department of Homeland Security must produce. HSPD 9 directs EPA to develop a surveillance and monitoring program to provide early warning in the event of a terrorist attack using diseases, pests, or poisonous agents. EPA is also charged, under HSPD 9, with developing a nationwide laboratory network to support the routine monitoring and response requirements of the surveillance program. HSPD 10 assigns additional responsibilities to EPA for decontamination efforts.

To obtain information for our analysis, we conducted a three-phase, Web-based survey of 43 experts on drinking water security. In identifying these experts, we sought to achieve balance in terms of area of expertise (i.e., state and local emergency response, engineering, epidemiology, public policy, security and defense, drinking water treatment, risk assessment and modeling, law enforcement, water infrastructure, resource economics, bioterrorism, public health, and emergency and crisis management). In addition, we attempted to achieve participation by experts from key federal organizations, state and local agencies, industry and nonprofit organizations, and water utilities serving populations of varying sizes. To obtain information from the expert panel, we employed a modified version of the Delphi method. The Delphi method is a systematic process for obtaining individuals' views and seeking consensus among them, if possible, on a question or problem of interest. Since first developed by the RAND Corporation in the 1950s, the Delphi method has generally been implemented using face-to-face group discussions. For this study, however, we administered the method through the Internet. We conducted our work in accordance with generally accepted government auditing standards between July 2002 and August 2003.

Experts Identified Key Vulnerabilities That Could Compromise Drinking Water Systems' Security

Our panel of experts identified several key physical assets of drinking water systems as the most vulnerable to intentional attack. In general, their observations were similar to those of public and private organizations that have assessed the vulnerability of these systems to terrorist attacks, including the National Academy of Sciences, Sandia National Laboratories, and key industry associations. In particular, as shown in figure 3, nearly 75 percent of the experts (32 of 43) identified the distribution system or its components as among the top vulnerabilities of drinking water systems. Experts also identified overarching issues compromising how well these assets are protected. Chief among these issues are (1) a lack of redundancy in vital systems, which increases the likelihood that an attack could render a system inoperable; and (2) the difficulty many systems face in understanding the nature of the threats to which they are exposed.

I would first like to discuss the distribution system, since it was cited most frequently as a key vulnerability by our panelists. The distribution system delivers drinking water primarily through a network of underground pipes to homes, businesses, and other customers. While the distribution systems of small drinking water utilities may be relatively simple, larger systems serving major metropolitan areas can be extremely complex. One such system, for example, measures water use through 670,000 metered service connections, and distributes treated water through nearly 7,100 miles of water mains that range from 2 inches to 10 feet in diameter. In addition to these pipelines and connections, other key distribution system components typically include numerous pumping stations, treated water storage tanks, and fire hydrants.

In highlighting the vulnerability of distribution systems, our panelists most often cited their accessibility at so many points. One expert, for example, cited the difficulty in preventing the introduction of a contaminant into the distribution system from inside a building "regardless of how much time, money, or effort we spend protecting public facilities." Experts also noted that since the water in the distribution system has already been treated and is on the way to the consumer, the distribution of a chemical, biological, or radiological agent in such a manner would be virtually undetectable until it was too late to prevent harm. While research on the fate and transport of contaminants within water treatment plants and distribution systems is under way, according to one expert, limited technologies are readily available that can detect a wide range of contaminants once treated water is released through the distribution system for public use.

Several other components, though not considered as critical as the distribution system, were still the subject of concern. Nearly half the experts (20 of 43) identified source water as among drinking water systems' top vulnerabilities. One expert noted, for example, that "because of the vast areas covered by watersheds and reservoirs, it is difficult to maintain security and prevent intentional or accidental releases of materials that could have an adverse impact on water quality." Yet some experts cited factors that mitigate the risks associated with source water, including (1) the source water typically involves a large volume of water, which in many cases could dilute the potency of contaminants; (2) the length of time (days or even weeks) that it typically takes for source water to reach consumers; and (3) the source water will go through a treatment process in which many contaminants are removed.

Also cited as vulnerabilities were the sophisticated computer systems that drinking water utilities have come to rely upon to manage key functions. These Supervisory Control and Data Acquisition (SCADA) systems allow operators to monitor and control processes throughout their drinking water systems. Although SCADA systems have improved water utilities' efficiency and reduced costs, almost half of the experts on our panel (19 of 43) identified them as among these utilities' top vulnerabilities.

Thirteen of the 43 experts identified treatment chemicals, particularly chlorine used for disinfection, as among utilities' top vulnerabilities. Experts cited the inherent danger of storing large cylinders of a chemical on site, noting that their destruction could release toxic gases in densely populated areas. Some noted, however, that this risk has been alleviated by utilities that have chosen to use the more stable liquid form of chlorine instead of the more vulnerable compressed gas canisters that have traditionally been used.

Finally, experts identified overarching issues that compromise the integrity of multiple physical assets, or even the entire drinking water system. Among these is the lack of redundancy among vital systems. Many drinking water systems are "linear"—that is, they have single transmission lines leading into the treatment facility and single pumping stations along the system, and often use a single computer operating system. They also depend on the electric grid, transportation systems, and single sources of raw materials (e.g., treatment chemicals). Many experts expressed concern that problems at any of these "single points of failure" could render a system inoperable unless redundant systems are in place. Experts also cited the lack of sufficient information to understand the most significant threats confronting individual utilities. According to the American Water Works Association, assessments of the most credible threats facing a utility should be based on knowledge of the "threat profile" in its specific area, including information about past events that could shed light on future risks. Experts noted, however, that such information has been difficult for utilities to obtain. One expert suggested that the intelligence community needs to develop better threat information and share it with the water sector.

Experts' Views on the Allocation and Distribution of Federal Funds

Many drinking water utilities have been financing at least some of their security upgrades by passing along the costs to their customers through rate increases. Given the cost of these upgrades, however, the utility industry is also asking that the taxpayer shoulder some of the burden through the appropriations process. Should Congress and the administration agree to this request, they will need to address key issues concerning who should receive the funds and how they should be distributed. With this in mind, we asked our panel of experts to focus on the following key questions: (1) To what extent should utilities' vulnerability and risk assessment information be considered in making allocation decisions? (2) What types of utilities should receive funding priority? and (3) What are the most effective mechanisms for directing these funds to recipients?

Regarding the first of these questions, about 90 percent of the experts (39 of 43) agreed "strongly" or "somewhat" that funds should be allocated on the basis of vulnerability assessment information, with some citing the vulnerability assessments (VAs) required by the Bioterrorism Act as the best available source of this information. Several experts, however, pointed to a number of complicating factors. Perhaps the most significant constraint is the Bioterrorism Act's provision precluding the disclosure of any information that is "derived" from vulnerability assessments submitted to EPA. The provision protects sensitive information about each utility's vulnerabilities from individuals who may then use the information to harm the utility. Hence, the law specifies that only individuals designated by the EPA Administrator may have access to the assessments and related information. Yet, according to many of the experts, even those individuals may face constraints in using the information. They may have difficulty, for example, in citing vulnerability assessments to support decisions on allocating security-related funds among utilities, as well as decisions concerning research priorities and guidance documents. Others cited an inherent dilemma affecting any effort to set priorities for funding decisions based on the greatest risk—whatever does not receive attention becomes a more likely target.

Regarding the second question concerning the types of utilities that should receive funding priority, 93 percent of the experts (40 of 43) indicated that utilities serving high-density population areas should receive a high or the highest priority in funding (See figure 4.). Fifty-five percent deemed this criterion as the highest priority. Most shared the view of one expert who noted that directing limited resources to protect the greatest number of people is a common factor when setting funding priorities. Experts also assigned high priority to utilities serving critical assets, such as national icons representing the American image, military bases, and key government, academic, and cultural institutions.

At the other end of the spectrum, only about 5 percent of the experts (2 of 43) stated that utilities serving rural or isolated populations should receive a high or highest priority for federal funding. These two panelists commented that such facilities are least able to afford security enhancements and are therefore in greatest need of federal support. Importantly, the relatively small percentage of experts advocating priority for smaller systems may not fully reflect the concern among many of the experts for the safety of these utilities. For example, several who supported higher priority for utilities serving high-density populations cautioned that while problems at a large utility will put more people at risk, utilities serving small population areas may be more vulnerable because of weaker treatment capabilities, fewer highly trained operators, and more limited resources.

Regarding the mechanisms for distributing federal funds, 86 percent of the experts (37 of 43) indicated that direct grants would be "somewhat" or "very" effective in allocating federal funds (See figure 5.) One expert cited EPA's distribution of direct security-related grant funds in 2002 to larger systems to perform their VAs as a successful initiative. Importantly, 74 percent also supported a matching requirement for such grants as somewhat or very effective. One expert pointed out that such a requirement would effectively leverage limited federal dollars, thereby providing greater incentive to participate.

The Drinking Water State Revolving Fund (DWSRF) received somewhat less support as a mechanism for funding security enhancements. About half of the experts

(22 of 43) indicated that the fund would be somewhat or very effective in distributing federal funds, but less than 10 percent indicated that it would be very effective.[1] One expert cautioned that the DWSRF should be used only if a process were established that separated funding for security-related needs from other infrastructure needs. Others stated that as a funding mechanism, the DWSRF would not be as practical as other mechanisms for funding improvements requiring immediate attention, but would instead be better suited for longer-term improvements.

Activities Experts Identified as the Most Deserving of Federal Support

When experts were asked to identify specific security-enhancing activities most deserving of federal support, their responses generally fell into three categories: (1) *physical and technological upgrades* to improve security and research to develop technologies to prevent, detect, or respond to an attack, (2) *education and training* to support, among other things, simulation exercises to provide responders with experience in carrying out emergency response plans, and specialized training of utility security staff; and (3) *strengthening key relationships* between water utilities and other agencies that may have key roles in an emergency response, such as public health agencies, law enforcement agencies, and neighboring drinking water systems.

As illustrated in figure 6, specific activities to enhance physical security and support technological improvements generally fell into nine subcategories. Of these, the development of "near real-time monitoring technologies," capable of providing near real-time data for a wide array of potentially harmful water constituents, received far more support for federal funding than any other subcategory—over 93 percent of the experts (40 of 43) rated this subcategory as deserving at least a high priority for federal funding. More significantly, almost 70 percent (30 of 43) rated it the highest priority—far surpassing the rating of any other category. These technologies were cited as critical in efforts to quickly detect contamination events, minimize their impact, and restore systems after an event has passed. The experts' views were consistent with those of the National Academies of Science, which in a 2002 report highlighted the need for improved monitoring technologies as one of four highest-priority areas for drinking water research and development.[2] The report noted that such technologies differ significantly from those currently used for conventional water quality monitoring, stating further that sensors are needed for "better, cheaper, and faster sensing of chemical and biological contaminants."

In addition to real-time monitoring technologies, the experts voiced strong support for (1) increasing laboratories' capacity to deal with spikes in demand caused by chemical, biological, or radiological contamination of water supplies, and (2) "hardening" the physical assets of drinking water facilities through improvements such as adding or repairing fences, locks, lighting systems, and cameras and other surveillance equipment. Regarding the latter of these two, however, some experts cited inherent limitations in attempting to comprehensively harden a drinking water facility's assets. In particular, they noted in particular that, unlike nuclear power or chemical plants, a drinking water system's assets are spread over large geographic areas, particularly the source water and distribution systems.

Regarding efforts to improve education and training, over 90 percent of the experts (39 of 43) indicated that improved technical training for security-related personnel warrants at least a high priority for federal funding. (See figure 7.) Over 55 percent (24 of 43) indicating that it deserved the highest priority. To a lesser extent, experts supported general training for other utility personnel to increase their awareness of security issues. The panelists also underscored the importance of conducting regional simulation exercises to test emergency response plans, with more than 88 percent (38 of 43) rating this as a high or highest priority for federal funding. Such exercises are intended to provide utility and other personnel with the training and experience needed both to perform their individual roles in an emergency and to coordinate these roles with other responders. Finally, about half the experts assigned at least a high priority to supporting multidisciplinary consulting teams ("Red Teams"), comprising individuals with a wide array of backgrounds, to provide independent analyses of utilities' vulnerabilities.

---

[1] The DWSRF program provides federal grant funds to states, which in turn allow the states to help public water systems in their efforts to protect public health and ensure their compliance with the Safe Drinking Water Act. States may use the funds to provide loans to public water systems, and may reserve a portion of their grants to finance other projects that protect sources of drinking water and enhance the technical, financial, and managerial capacity of public water systems.

[2] *Making the Nation Safer: The Role of Science and Technology in Countering Terrorism*, p. 250. The National Research Council of the National Academies. (Washington, D.C.: The National Academies Press, 2002).

As illustrated in figure 8, experts also cited the need to improve cooperation and coordination between drinking water utilities and certain other organizations as key to improving utilities' security. Among the organizations most often identified as critical to this effort are public health and law enforcement agencies, which have data that can help utilities better understand their vulnerabilities and respond to emergencies. In addition, the experts cited the value of utilities' developing mutual aid arrangements with neighboring utilities. Such arrangements sometimes include, for example, sharing back-up power systems or other critical equipment. One expert described an arrangement in the San Francisco Bay Area—the Bay Area Security Information Collaborative (BASIC)—in which eight utilities meet regularly to address security-related topics. Finally, over 90 percent of the experts (39 of 43) rated the development of common protocols among drinking water utilities to monitor drinking water threats as warranting a high or highest priority for federal funding. Drinking water utilities vary widely in how they perceive threats and detect contamination, in large part because few common protocols exist that would help promote a more consistent approach toward these critical functions. Some experts noted, in particular, the need for protocols to guide the identification, sampling, and analysis of contaminants.

Observations

In 2002, EPA's Strategic Plan on Homeland Security set forth the goal of significantly reducing unacceptable security risks at water utilities across the country by completing appropriate vulnerability assessments; designing security enhancement plans; developing emergency response plans; and implementing security enhancements. The plan further committed to providing federal resources to help accomplish these goals as funds are appropriated.

Key judgments about which recipients should get funding priority, and how those funds should be spent, will have to be made in the face of great uncertainty about the likely targets of attacks, the nature of attacks (whether physical, cyber, chemical, biological, or radiological), and the timing of attacks. The experts on our panel have had to consider these uncertainties in developing their own judgments about these issues. These judgments, while not unanimous on all matters, suggested a high degree of consensus on a number of key issues.

We recognize that such sensitive decisions must ultimately take into account political, equity, and other considerations. But we believe they should also consider the judgments of the nation's most experienced individuals regarding these matters, such as those included on our panel. It is in this context that we offer the results presented in this testimony as information for Congress and the administration to consider as they seek the best way to use limited financial resources to reduce threats to the nation's drinking water supply.

Mr. Chairman, this completes my prepared statement. I would be happy to respond to any questions you or other Members of this Subcommittee may have.

Figure 1: Key Components of a Typical Drinking Water System

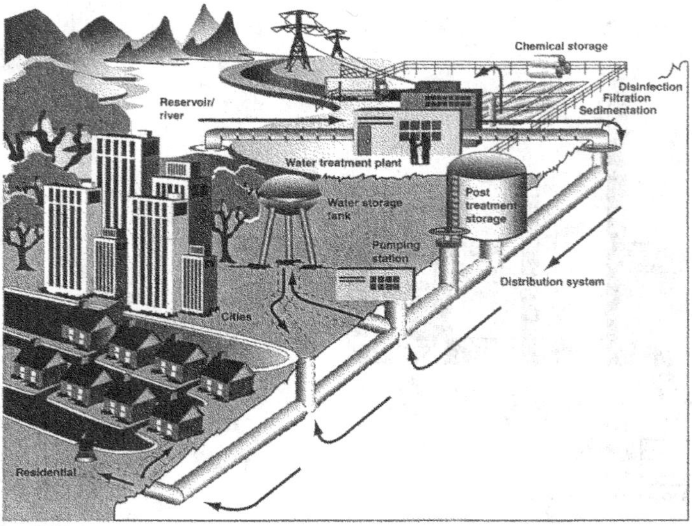

Source: GAO.

Figure 2: Number of Drinking Water Systems That Serve Various Populations

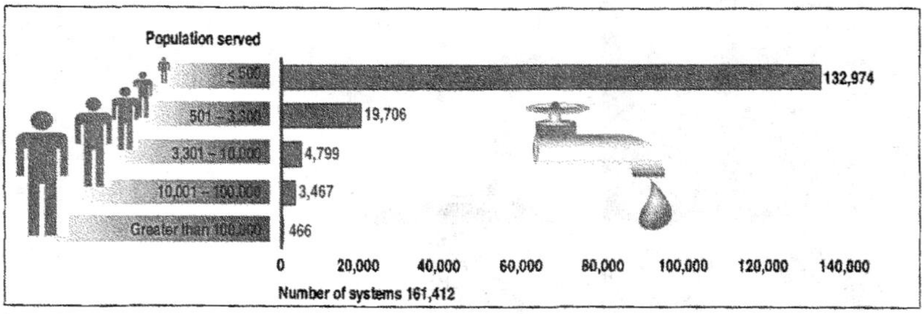

Source: GAO.

Figure 3: Key Vulnerabilities Identified As Compromising Drinking Water Systems' Security

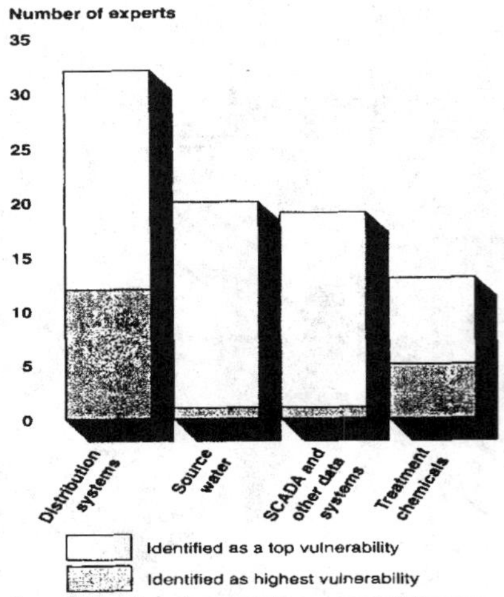

Source: GAO analysis of expert panel's responses to GAO survey.

Figure 4: Experts' Views on Which Types of Water Utilities Should Receive Priority for Federal Funds

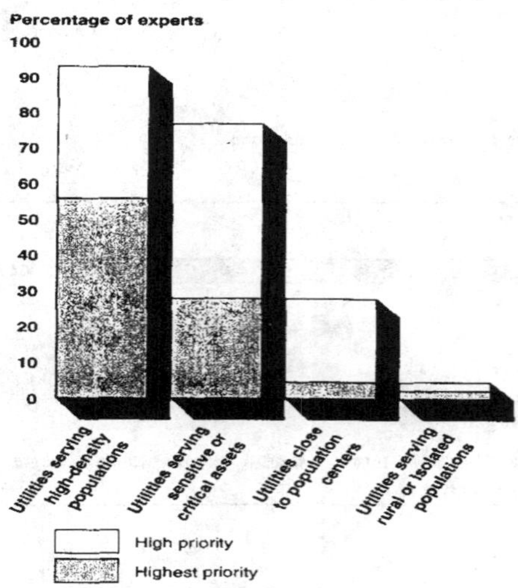

Source: GAO analysis of expert panel's responses to GAO survey.

Figure 5: Recommended Approaches to Distribute Federal Funds

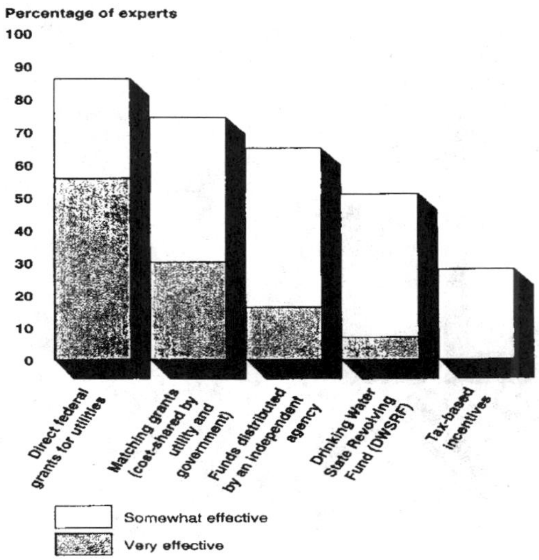

Source: GAO analysis of expert panel's responses to GAO survey.

Figure 6: Activities Identified by Expert Panel to Enhance Physical Security and Support Technological Improvements

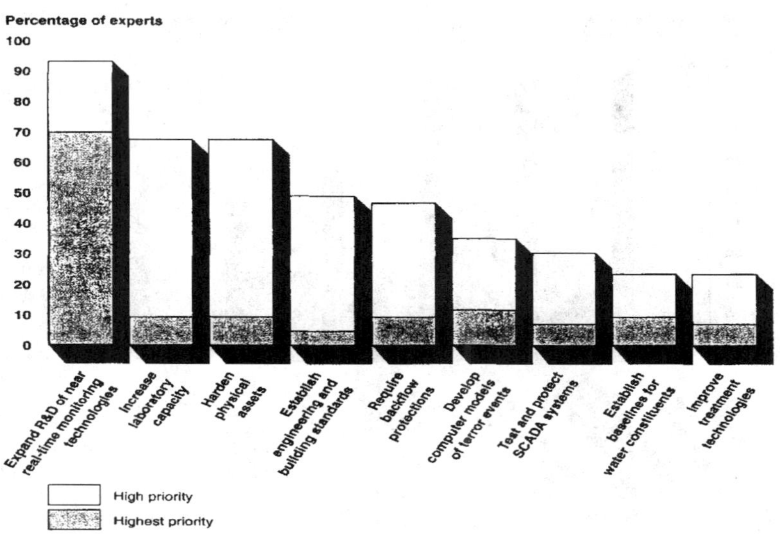

Source: GAO analysis of expert panel's responses to GAO survey.

Figure 7: Activities Identified by Experts to Improve Education and Training

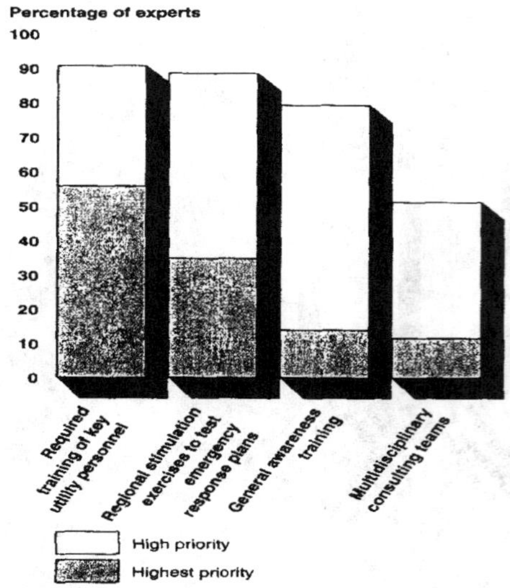

Source: GAO analysis of expert panel's responses to GAO survey.

Figure 8: Activities Identified by Experts to Strengthen Relationships Among Agencies and Utilities

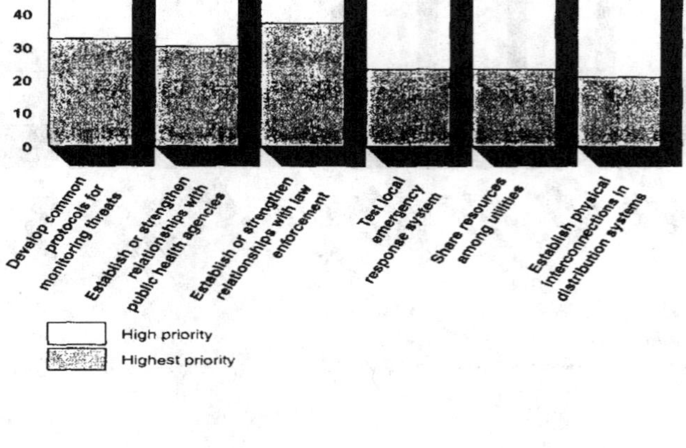

Mr. GILLMOR. Thank you very much, Mr. Stephenson.

The Chair will recognize himself for some questions.

First question to Mr. Grumbles, or you could jump in, too, if you wanted, Mr. Stephenson, is could you give us some kind of indication as the amount of poison or contaminant it would take to have widespread impact on a system? I think there is, you know, the movie version where you get some bad guy with a vial, he dumps it in a reservoir and he poisons a city. Other people have said to really cause any widespread harm you would have to be dumping contaminants equal to several tanks, trucks full.

Could you just give us some kind of ballpark indication of how much contaminant it would take to poison a system?

Mr. STEPHENSON. Well, I will make a comment before Ben does. But that is, in fact, why the distribution system was cited as the greatest vulnerability. Source water, such as a reservoir, is pretreatment, and is not a very effective way to contaminate drinking water.

However, when is the last time you saw a truck backed up to a fire hydrant and assumed he was taking water out of the system? He could just as easily be putting a contaminant into the system. Since it is post treatment, it goes directly to the homes or businesses from there. And that is why I think our experts cited that as a very high vulnerability.

Mr. GRUMBLES. Mr. Chairman, I think the question you asked is on the minds of a lot of people, and the simple basic answer is that it truly does depend—it depends on the contaminant and the situation. So there is—no real short answer to that. It just largely varies, based on the contaminant involved and the nature of placing the contaminant within the system.

Mr. GILLMOR. To follow up on that, during the hearing in the House Science Committee in 2001—and I think actually you were working on the staff there at the time—EPA was working on a state of knowledge report with DOD, Centers for Disease Control, FBI, Food and Drug Administration, the Office of Homeland Security, to compile and assess known information about biological, chemical and radiological contaminants, as well as a detection technology.

Could you tell us what the status is of that working group, and is EPA still working to find out about new contaminants?

Mr. GRUMBLES. Yes, sir. Based on the findings of an interagency work group that we convened to address the state of the knowledge—this is the report that you referenced—we include recommendations on vulnerabilities of water systems that should be considered, along with the potential mitigation actions in our baseline threat document.

Our baseline threat document is a critically important document that we provided to utilities back in 2002, to help assist them in vulnerability assessments and to prepare an emergency response plan. So what we are currently doing is undertaking analyses to fill in gaps in the knowledge and working with other agencies as well to do so.

Developing analytical methods for some contaminants, for determining the effectiveness of disinfection practices for particular contaminants. That is some of the examples of the things we are doing. So I appreciate the chance to talk a little bit about that state of the knowledge report and how we have been following up.

Mr. GILLMOR. So basically, you have been disseminating some of that information through the assessment process; is that correct?

Mr. GRUMBLES. That is correct, providing it to the utilities to help inform them. I guess it is important to clarify as well. We need to provide the information to utilities as they were tasked with developing their vulnerability assessments. We did indeed flag concerns about terrorism.

It wasn't just the pre9/11 world, it was post 9/11. So it was very important to clarify. We included that information in terms of the baseline threat document.

But the state of the knowledge report also fed into that effort to provide guidance.

Mr. GILLMOR. A quick question for Mr. Stephenson. Recognizing that rural water systems are at least able to afford security enhancements, but also recognizing that even in your study, your experts suggested that utilities serving high-density populations should receive the highest priority in funding.

Do you have any suggestions as to how we ought to allocate Federal funding and in that respect, did your study have a significant amount of experts representing rural communities?

Mr. STEPHENSON. I believe we had a member of the Rural Water Association on it. But obviously the high density populations—the 466 large systems that serve over 50 percent of the population always came up as a higher priority—even though we tried to get balanced representation across our 43 experts. As you know, the Bioterrorism Act addresses systems bigger than 3,300 but even at that suggests that EPA provides guidance to the smaller utilities, and I believe there are over 160,000 facilities that serve less than 500. So it becomes economies of scale, I guess as to how you could best provide funding for those small public systems.

Mr. GRUMBLES. Mr. Chairman, I would just like to add that while the priorities—and the focuses and the timeframes in the Bioterrorism Act focused on the larger systems—we have not lost sight of the fact that thousands and thousands of smaller systems in the country should be doing their part as well. EPA has provided—over the last several years—about 25 percent of its budget for the water security efforts to the small systems, $36 million for training and technical assistance and working with small systems through workshops and indirect assistance with rural water circuit riders.

Mr. GILLMOR. Thank you.

The gentlelady from California.

Ms. SOLIS. Thank you, Mr. Chairman. This is for Mr. Grumbles from EPA. The Inspector General suggested that the EPA needs to analyze the quality of the vulnerability assessments submitted by large utilities to determine whether they adequately address the threats envisioned by the Bioterrorism Act. Has the EPA analyzed the quality of all the vulnerability assessments of the 350 largest systems that served over 116 million Americans?

Mr. GRUMBLES. Congresswoman, I really appreciate the question. In my conversations with the Inspector General and in taking to heart recommendations she has, sometimes we don't always agree with them, but they are always helpful to see where the Inspector General might have information.

On this particular issue, what we agreed to do was to convene a prestigious subgroup of the National Drinking Water Advisory Council, the Water Security Working Group, and we have specifically tasked them with developing measures to gauge "the quality or the effectiveness of the plans."

Ms. SOLIS. How many have been analyzed? Do you have a number?

Mr. GRUMBLES. How many have been?

Ms. SOLIS. Of the largest 350 that have actually been analyzed by EPA?

Mr. GRUMBLES. Well, we have received all of the vulnerability assessments.

Ms. SOLIS. Right. But that doesn't mean that you have analyzed them. Are they?

Mr. GRUMBLES. Well, we follow the—a couple of points. First of all, we have very specific framework for reviewing the quality of the vulnerability assessments as laid out in the statute. Our job and interpretation of the statute has been that EPA reviews them to insure compliance with the basic requirements, provisions of the law—as is the intent of Congress—and so we have reviewed all of the vulnerability assessments on the quality component.

What we have specifically asked is to get this independent group that includes experts from various sectors, governmental, non-governmental to look at that issue and to develop measures for effectiveness to help address that question of what is the quality of the vulnerability assessments.

Ms. SOLIS. So that can vary depending on whatever report you can get from one of these 350 large systems? I mean, this is very—this is somewhat, very subjective.

Mr. GRUMBLES. Well, a couple of things again. I think that one of the things that, the Drinking Water Advisory Council, their water security working group, is not specifically reviewing each of the vulnerability assessments themselves, that is an important point. And the way the law is currently written, I don't think that would be legal, unless we——

Ms. SOLIS. Why would that not be legal?

Mr. GRUMBLES. Unless we designated each and every one of those.

Ms. SOLIS. I guess what I am trying to understand, if you have 350 large systems that you are supposed to be collecting data for, and you want to try to get some standard or criteria, if they are all needing—or what we have set out that they should need, and you have an expert group looking at that, it doesn't really give me a sense that we have some uniformity here. I mean, it could vary. You could get different feedback from different parts of the——

Mr. GRUMBLES. I don't think we have, there is—not a real disagreement here. I think I need to communicate more clearly.

We recognize that it is important not to have just a subjective—I mean, basically what the Inspector General was getting at, I believe, was what is the overall quality of these vulnerability assessments?

Ms. SOLIS. But they have questioned that. They have questioned your measurement of that.

Mr. GRUMBLES. Right. And we working with them said, you know, it would be good to get some helpful criteria to define what is an effective security program. And that specifically is what the Drinking Water Advisory Council is tasked to come up with to help shed some light to share and share with us what that is.

So I think——

Ms. SOLIS. My understanding is that you do have clear authority to do assessments for each of these vulnerability assessment studies. That is my understanding. You have just said something different earlier.

Mr. GRUMBLES. No. We do specifically review each of the vulnerability assessments.

Ms. SOLIS. But you haven't. You haven't done all of them?

Mr. GRUMBLES. No. We have. We have them all. We have received them all within the agency.

Ms. SOLIS. Right. And they have all been analyzed and assessed by the EPA.

Mr. GRUMBLES. My staff is informing me that the large ones, the 400-plus vulnerability assessments that we have received, we have reviewed.

Ms. SOLIS. Can I have that? I would like to have that in writing——

Mr. GRUMBLES. Sure.

Ms. SOLIS. [continuing] for this committee. And if you can guarantee that as well.

My time is almost up. I just have a question with respect to when, say, a water facility submits their plan and after you find that there might be some questions or issues about their plan with respect to—I don't want to say terrorism—but, say, maybe disgruntled employees that may disrupt the facility. How do you separate that out from looking at plants for addressing terrorism?

Mr. GRUMBLES. Let me make sure I understand. Separating vandalism from terrorism?

Ms. SOLIS. My understanding is that there has been a lot of reporting of that in these plans, and there hasn't been enough?

Mr. GRUMBLES. In the emergency response plans or vulnerability assessment?

Ms. SOLIS. Vulnerability assessment.

Mr. GRUMBLES. Am I allowed to talk about—you know, I would welcome the opportunity to talk with you on—to the extent I can, the specifics of the vulnerability assessments, but, I think, not in a public forum.

Ms. SOLIS. Okay.

Mr. GRUMBLES. I would like to emphasize, and clarify what I have said with respect to the Inspector General, we have worked with the Inspector General, designated several of their people to actually look at and review the vulnerability assessments. And so if I, if that was not clear, I wanted to make sure that that was clear.

Mr. GILLMOR. I will come back, I guess, to that question.

Mr. GRUMBLES. And that was after, after they gave us the report, we said we will designate you and you can review, actually review the vulnerability assessments. So. I think we are working with them—and we want to particularly also get the working group from the National Drinking Water Advisory Council to have some objective criteria as to what is a successful program.

Ms. SOLIS. I guess in the report that I am seeing right in front of me, right now, you believe that you can analyze information in vulnerability assessments because this would violate the Public Health Security Safety and Bioterrorism Preparedness and Response Act. That was your response. However, the council says that EPA does, in fact, have the authority as well as the responsibility to collect and analyze necessary information on these sources. That is what the inspector general said.

Mr. GRUMBLES. And I believe that we, through Congress' leadership in appropriating funds for grants for vulnerability assessments—it is also our responsibility to ensure that the purposes of the grant are carried out; and so the basic requirements that are in those statutes about vulnerability assessments are done. So that as we have gotten the large vulnerability assessments and reviewed them, we have specifically looked at those factors, taken that into account.

And we welcome the inspector general's comments and the National Drinking Water Advisory Council's objective criteria that they will use for quality.

Mr. GILLMOR. The gentlelady's time has expired.

Just as a point of information, the limitations that Mr. Grumbles was talking about were, I think, in section 1435 of the conference report of the Bioterrorism Act as to what type of thing they can do.

The gentleman from Michigan, Mr. Rogers.

Mr. ROGERS. I will pass.

Mr. GILLMOR. The gentleman passes.

The other gentleman from the northern peninsula.

Mr. STUPAK. Thank you, Mr. Chairman.

Mr. Grumbles, in this legislation you got $160 million to do these assessments, these vulnerability assessments. And the way I read this report, EPA has not done a very good job.

While you have done a good job of getting the reports in, the vulnerability assessments are really pre-9/11. In other words, these assessments, because of lack of leadership from the EPA, more utilities focused on vandals, criminals, and disgruntled employees in their vulnerability assessment.

Contractors further stated—these contractors that Mr. Stephenson talked about, further stated that EPA has not provided utilities the intelligence data or threat information required to justify the security upgrades necessary to defend against terrorism. While the terrorist attacks of 9/11 and subsequent passage of the Bioterrorism Act served as a catalyst for the vulnerability assessments, limited threat information provided by the EPA resulted in utilities subjectively designing their assessments around pre-9/11 threats.

After filtering threat information through the RAMW methodology, most of the water security experts we interviewed who were familiar with vulnerability assessments concluded that the only threats utilities could realistically address were those they encountered before 9/11, being the vandals, the criminals and disgruntled employees. One utility representative we interviewed said that the contractor they hired to conduct their vulnerability assessment discouraged them from addressing higher threat levels like terrorism.

So, in answer to Ms. Solis' questions and that, while you have a lot of paperwork to submit to the EPA, it doesn't meet the guidelines put forth by the Bioterrorism Act, and it is really not a question of money because, you know, $160 million should have at least given us some kind of ideas, not what happened before 9/11.

The reason why you had the Bioterrorism Act was because of 9/11, and it seems like we have missed the whole point here because of lack of guidance from the EPA.

Mr. GRUMBLES. Congressman, I would say a couple of things. One, I would respectfully—respectfully would disagree with the characterization in the sense that when we put together the baseline threat document and when we put additional guidance forward with utilities, we did emphasize terrorism and terrorist attacks.

I would also say that based on our customer surveys, the information we have gotten from our customers, the utilities, we have gotten a large sense of satisfaction in terms of the guidance and information that was provided.

The last point I would make, Congressman, is that while we feel that we have provided guidance, helpful guidance, to utilities in developing vulnerability assessments this first round, these initial vulnerability assessments are not the be all and end all. I fully——

Mr. STUPAK. I would hope not, because you haven't even addressed terrorism according to this report. And this isn't the Office of Inspector General; this is from your own internal documents. This is your own reports.

Mr. GRUMBLES. I think that the vulnerability assessments are viewed as a step forward. I think that they continue to, and will, improve. And—they need to be living documents, and they will only improve.

And I think utilities have done a good job in this first round. We are in a new era after the Bioterrorism Act, and I think they have done a good job, and our job is to provide them additional information.

Mr. STUPAK. Your own document says—from Jeffrey K. Harris, Director of Program Evaluations, Cross-Media Issues, was to Tracy Meehan, Assistant Administrator for Office of Water, certainly doesn't say that. And if you look at the IG report, it says—let me quote here on page 5:

"One of the security experts we interviewed stated the EPA did not provide adequate threat information. Officials at the Sandia National Laboratory stated that the EPA threat guidance missed the mark because EPA did not set minimum threat levels against which utilities need to assess their vulnerabilities."

If you don't have any standard, I guess you could call anything a success because you have no standard to judge it against. And that is where we think the leadership is lacking from the EPA.

Mr. GRUMBLES. Congressman, do you know if the inspector general had reviewed any of the vulnerability assessments when that statement was made?

See, my information——

Mr. STUPAK. Well, that is why we want the inspector general here. If he is not here, we can't answer it. You can't answer it. I can't answer it.

So let me ask you this one: How about you? Has the EPA exercised its authority and required any drinking water utilities to take corrective actions to address vulnerabilities to terrorist acts or other intentional acts? Have you, EPA, exercised your authority requiring them to do anything to take corrective action, other than submit these plans?

Mr. GRUMBLES. I guess my point——

Mr. STUPAK. No, no, just a yes or no.

Come on now, corrective action or not. Did you guys direct anyone to take corrective action or not?

Mr. GRUMBLES. Enforcement action?

Take any enforcement—I don't know.

Mr. STUPAK. You have the authority and are required that if there is a lack of security at these water utility places, you have the right and the authority to require them to take corrective action. Have you done that, in looking at these plans, since you had 400 sitting in your office from the large utilities?

Mr. GRUMBLES. We take our responsibilities and authority seriously. And I am not sure that we have the authority to take an enforcement action in that situation.

I can assure you——

Mr. STUPAK. Okay. Whether you need the authority, did EPA do anything, whether you had the authority or not?

Let us pretend you had it for a minute, okay? Did you take any corrective action, or are all these plans, all 400, just perfect?

Mr. GRUMBLES. If we had the authority, then our first step would be to ensure compliance assurance. And then if the community didn't take those steps, then we would take an enforcement action.

Mr. STUPAK. So you haven't taken any enforcement action yet?

Mr. GRUMBLES. Well, we think that we have exercised the current authorities that we have, current legal authorities. And I would emphasize, Mr. Chairman, that the whole—the underlying basis for progress here is partnership with the communities.

Mr. STUPAK. Well, you know, under authority—I am looking at a letter here April 22, 2002. It is from Christy Todd Whitman, EPA Director, to John Dingell. On page 2 of that letter it says, "The language contained in 3448"—that is the bioterrorism bill—"amending the Safe Drinking Water Act, section 1431, provides EPA with adequate authority to respond in situations involving significant vulnerability."

So that is why I asked my question, since you have the authority and are required. So did you take any action to address these significant——

Mr. GRUMBLES. I appreciate your clarifying that because that is the provision in the statute that deals with imminent and substantial endangerment. And I am not—I don't—and fortunately, I don't think we have had any situations where we have exercised that rare authority to step in in the context of a vulnerability assessment or an emergency response plan.

Mr. STUPAK. I am out of time, but I will keep going if you let me.

Mr. GILLMOR. How about we come back for another round?

Mr. STUPAK. Sure.

Mr. GILLMOR. Mr. Terry apparently got tired of waiting, so we will recognize him when he comes back, although he did have a couple of questions.

In fact, I know what he was going to ask. I might ask one of those questions on his behalf; and that is, one of the bigger concerns is making sure that EPA is not ignoring small water systems in order to focus solely on the largest drinking water systems. It was his understanding that EPA had been using money to provide the trainer grants, to provide a number of environmental professionals to give training and technical assistance to water systems serving fewer than 50,000 people.

And then his question was: "What is the status of this program?"

Mr. GRUMBLES. Yes, sir.

Mr. GILLMOR. And what other actions is EPA taking to help smaller water systems?

And the third part of that was, what fraction of the drinking water security budget is being spent on smaller systems?

Mr. GRUMBLES. Since 2002, what EPA has done is what—we have provided $36 million for training and technical assistance for the small water systems, those less than 50,000, under the terms of the Bioterrorism Act. This is approximately 25 percent of the budget 2002 through 2004.

We have used a multi-pronged approach, Mr. Chairman, to try to reach the nearly 8,000 systems that are—those less than 50,000 that are required to undertake vulnerability assessments and develop or revise emergency response plans. Besides training the trainers, Mr. Chairman, we provided direct assistance to trainers such as the rural water circuit riders.

So, as a result, the number I have is that more than 88 percent of the small systems have met the deadline for submitting the vulnerability assessments. So, again, while the focus, I think, of the drafters and of the Nation also is on timeframes for the large metropolitan areas, we certainly recognize the importance of getting out assistance and ensuring that the smaller systems throughout the country are also assessing and developing emergency response plans and getting the information they need to secure their systems.

Mr. GILLMOR. Then also—although not as good looking as Mr. Terry, I am standing in for him.

Mr. Stephenson, how helpful will real-time monitoring technologies, capability of providing near real-time data for a wide array of potentially harmful contaminants be in addressing security issues? And why do you think this technology received the most support for Federal funding?

Mr. STEPHENSON. I think the experts felt that because the distribution system was the most vulnerable, these detection and monitoring capabilities would be placed in the distribution systems so that would give real-time information if there was a contaminant in the system. Facilities currently have no capacity to do this. So they felt a lot of research was needed in that area, more so than hardening or anything else; and I think that is why that cropped up as the highest priority from our experts.

Mr. GILLMOR. And Mr. Grumbles, how does EPA receive threat information? Does it come from FBI, DHS, or other parts of the Intelligence Community? And what is the extent of collaboration with DHS? And what are the procedures to make sure that information flows down the chain to water systems in a timely manner?

Mr. GRUMBLES. Well, Mr. Chairman, how do we get our information? We get it through a variety of sources, primarily DHS, FBI, CIA. We are working very closely with the Department of Homeland Security. While we are the sector-specific lead for the water infrastructure sector, we do report to them and we work very closely with them.

And we also within the agency have an Office of Emergency Preparedness, as well as an Office of Homeland Security, to help provide specific liaison to the other agencies throughout the Federal family of Homeland Security individuals.

Mr. GILLMOR. Thank you, Mr. Grumbles.

And we will go to another round of questions.

And, Ms. Solis.

Ms. SOLIS. Thank you, Mr. Chairman. I would like to request unanimous consent to also submit several editorial articles to support action to decrease threats at chemical facilities.

Mr. GILLMOR. Without objection.

[The information referred to follows:]

Editorial Pages Support Strong Action to Decrease Threats at Chemical Facilities across the Country

(July 2002 – September 2004)

National

USA Today, "Remembering 9/11- The task," September 10, 2004

Washington Post, "Safeguarding Chemical Plants," April 07, 2003

Washington Post, "Seeking Chemical Safety," September 14, 2002

Alabama

Anniston Star, "Chemical security," June 07, 2004

California

Los Angeles Times, "Rights and the New Reality: Chemical Industry vs. Safety," September 18, 2002

Los Angeles Times, "RIGHTS AND THE NEW REALITY: Fine-Tuning Chemical Rules," July 28, 2002

Florida

Fort Lauderdale Sun-Sentinel, "2 Years Later, Threat Persists," September 11, 2003

Fort Lauderdale Sun-Sentinel, "Safeguard U.S. Chemical Plants," March 26, 2003

Fort Lauderdale Sun-Sentinel, "Safeguard U.S. Chemical Stocks," August 02, 2002

Melbourne Florida Today, "Protecting chemicals needs high priority," August 06, 2002

Sarasota Herald Tribune, "Hazardous Chemicals: Does industry's influence outweigh public's safety?," September 29, 2004

Sarasota Herald-Tribune, "Delayed reaction: Lax security at chemical plants: a disaster waiting to happen," August 02, 2004

St. Petersburg Times, "Unsecured soil," March 30, 2003

St. Petersburg Times, "Chemical plants at risk: The government should beef up its efforts to secure chemical plants as it has with other sites vulnerable to terrorist attack, and not hope for self-regulation," August 14, 2002

Georgia

Atlanta Journal-Constitution, "Elements of concern: Congress, Bush administration must act quickly to improve security at nation's chemical facilities," August 26, 2004

Atlanta Journal-Constitution, "Order chemical industry to fall in," April 09, 2003

Kansas

Wichita Eagle, "SECURE: Plants need protection," April 14, 2003

Louisiana

New Orleans Times-Picayune, "Securing chemical plants," July 10, 2003

Massachusetts

Boston Globe,."Deadly chemicals," June 02, 2004

Boston Globe, "Unprotected targets," September 21, 2003

Boston Globe, "Hazardous chemicals," May 09, 2003

Minnesota

Minneapolis Star Tribune, "After three years, U.S. is a long way from safe," September 12, 2004

New Hampshire

Keene Sentinel, "Nuclear concern: High time to wonder about safety," August 22, 2004

Keene Sentinel, "Chemical duty," April 17, 2003

New Jersey

Asbury Park Press, "Part 4: Terrorist target on reactor's back," June 23, 2004

Asbury Park Press, "Chemical plants still vulnerable," January 03, 2004

Asbury Park Press, "Improving security at chemical plants," August 15, 2003

Asbury Park Press, "Reducing the threat," April 29, 2003

Asbury Park Press, "Safeguarding chemical plants," March 31, 2003

Asbury Park Press, "Reducing threats from chemicals," July 30, 2002

Asbury Park Press, "Risk reduction, not secrecy," July 17, 2002

Press of Atlantic City, "CHEMICAL SECURITY BILL: Safety is priority," January 07, 2004

Bergen Record, "Chemical targets," December 28, 2003

Bergen Record, "Chemical alert," March 19, 2003

Bergen Record, "The need to make chemical facilities safer," February 25, 2003

Bergen Record, "Chemical-plant safety: Why security upgrades must be mandatory," September 18, 2002

Bergen Record, "Prime targets: Reducing the dangers from chemical plants," July 24, 2002

Newark Star-Ledger, "Thwarted on chemical threat," May 28, 2004

Newark Star-Ledger, "Securing chemical plants," November 22, 2003

Newark Star-Ledger, "Chemical plant security," April 03, 2003

Newark Star-Ledger, "Chemical plant security," March 19, 2003

Newark Star-Ledger, "Securing chemical plants," September 22, 2002

New York

Buffalo News, "A question of security: Stringent protection is needed at the nation's chemical plants," May 08, 2003

Buffalo News, "Abridging freedom of information," September 14, 2002

Long Island Newsday, "Chemical Warfare: ... can be waged by terrorists against U.S. chemical plants, so federal rules are needed," April 05, 2003

New York Times, "Chemical Security," May 05, 2003

New York Times, "Avoiding Chemical Catastrophe," April 01, 2003

North Carolina

Charlotte Observer, "Chemical safety: How to think positive thoughts about toxic clouds," October 09, 2003

Charlotte Observer, "Chemical terrorism: Industry, agencies unprepared to prevent or respond to it," June 09, 2003

Charlotte Observer, "Chemical safety: Security measures need to be broader and better now," April 17, 2003

Charlotte Observer, "Chemical danger: Security at U.S. plants still hasn't been addressed," April 04, 2003

Pennsylvania

Philadelphia Inquirer, "Chemical Plants: A clear and present danger," September 16, 2004

Philadelphia Inquirer, "Fiddling around: Why the delays on chemical plant security?," April 05, 2003

Philadelphia Inquirer, "Fools on the Hill: No excuse for feds to dawdle on tough security standards at chemical plants," March 05, 2003

Philadelphia Inquirer, "Support safety, senators: Pass the Corzine bill on chemical plants," September 18, 2002

Philadelphia Inquirer, "To heal and defend: As the first anniversary passes, looking to tasks the nation still faces," September 12, 2002

Philadelphia Inquirer, "Chemical weapons?: Strengthen safety and security at U.S. plants," August 13, 2002

South Carolina

Charleston Post & Courier, "Our continuing vulnerability," August 10, 2004

Texas

Dallas Morning News, "Terrorist Targets: U.S. must ensure chemical plants are safe," July 16, 2004

Houston Chronicle, "VOLATILE: Self-policing too puny to avoid chemical plant sabotage," September 26, 2002

Virginia

Roanoke Times, "Tend to the basics of homeland security: Shoring up essential precautions will help minimize chances of a terrorist nuclear attack," March 22, 2004

Roanoke Times, "Homeland security on the cheap," April 07, 2003

Roanoke Times, "For true homeland security, apply the law to chemical plants: The industry's lobbyists have sought to gain exemptions from oversight. The Senate should not consent," September 09, 2002

Wisconsin

Milwaukee Journal Sentinel, "Holes in homeland defense," February 22, 2003

Ms. SOLIS. Thank you.

Mr. Grumbles, going back to Los Angeles, we are a very large urban center there, and obviously the DWP is one of the largest water providers in Los Angeles. They have submitted their plan, a 5-year plan, and my understanding is that you have received that. But, looking at it, there are so many issues that just kind of beg to be answered.

One is that they have a budget problem with respect to employees there having to somewhat police and provide surveillance for their facilities. And overtime is a big issue because they have not come up with, say, installing electronic surveillance equipment.

What types of advice do you give to agencies like that to urge them or at least to direct more grant money so that they can accomplish this goal, knowing that they are faced with these—and L.A. isn't the only one. I am sure this is with a lot of other facilities.

How is it that you get back to them, and what is the timeframe if there are changes that you think could help or to modify their plan? What is it that you do to get back to them?

Mr. GRUMBLES. I—you mentioned L.A., and that is a perfect example of a city where there is so much at stake and where initiative has already occurred and they are moving out front. And then it translates into, they have developed plans and very specific milestones to try to increase the security of their system—how do they finance it and fund it, and how do they, you know, get there.

One of our jobs that we take extremely seriously is to provide not only the tools and the training, but the technical assistance, ways to find additional funding, and to also use some of the existing Federal funding that might be there, if not through the Department of Homeland Security, through EPA. We have some funding through the drinking water State revolving funds, and we issue guidance expressly for the purpose of going to States to help the cities use some of those funds that traditionally have been used for drinking water, maximum contaminant level compliance and drinking water treatment plants, and to use those funds for security-related enhancements.

Some of these areas, when you talk about overtime or increased O&M, those may not be eligible for assistance under that drinking water State revolving fund, and so there are other tools or financial assistance. Sometimes, as you and your constituents know better than anyone, ultimately it is the ratepayer, the customer, that is going to be paying more for enhanced security, just like they pay more for enhanced drinking water regulations that we issue.

Ms. SOLIS. At what point do you get back to the different water purveyors, especially the large ones, in terms of their plan, though? What do you do to go back and maybe review or even audit?

Mr. GRUMBLES. Are you talking about—when you say "a plan," are you talking about——

Ms. SOLIS. Vulnerability.

Mr. GRUMBLES. The vulnerability assessment?

Ms. SOLIS. Yes. If there are some questions, a red flag goes up or something, how quickly do you get back to them to let them know that you perceive there is a problem or an issue here?

Mr. GRUMBLES. I don't know a timeframe.

Ms. SOLIS. I mean, this is obviously very, very important. And you haven't set up any standard to get to do that?

Mr. GRUMBLES. We have a good dialog with the drinking water utility associations across the country. We meet regularly with them, provide them information.

Ms. SOLIS. That is not what I am asking.

Mr. GRUMBLES. Well, I think one of the mechanisms—probably the Region 9 office might be the closest EPA office to get back to them on some of the specifics of the questions they might have. But our Office of—our Water Security Division does provide information.

Some of the—Congresswoman, some of the venues that we would use would be through our workshops that we have with utilities and cities. We are very pleased that we are part of the funding and supporting the water—ISAC, Information Sharing and Analysis Center, which is a secure Web-based system that cities, large and small, utilities, once they get security pass words can use to get helpful information on some of their very real security-related issues.

Ms. SOLIS. But limited to the funds, very limited funds available, right, to make any changes?

Mr. GRUMBLES. They would certainly say that. And I would say, from a Federal EPA budget, funding is always a difficult challenge. And as we move into the implementation stage, it will continue to be a challenge. But we are taking that very seriously and looking at funding as a Federal partner with localities and States as we look into the next budget cycle.

Mr. GILLMOR. The gentlelady's time has expired.

The gentleman from Michigan.

Mr. STUPAK. Thank you, Mr. Chairman.

Mr. Grumbles, sort of getting back to where I left off, let me ask the question this way: Has the EPA requested any drinking water utility to take specific corrective action to address vulnerabilities to terrorist attacks?

Mr. GRUMBLES. I guess I should say I would like to make sure that I answer it correctly, and I probably need to get back to you for purposes of the record on that.

I am told no.

Mr. STUPAK. Okay. The large utilities were supposed to be done March 31, 2003, to submit their vulnerability assessments to you. That has been 18 months ago, and we haven't directed anyone to take any corrective actions. So the answer on that is no, right?

So then, if that is the case, you said in your opening statement that "We are smarter and we are safer from terrorist attack because of the work of the EPA." but with all seriousness, how is the public assured that the necessary security enhancements are being taken by their water utilities?

We have these assessments done; there has been no corrective action. How do we reassure the public?

Mr. GRUMBLES. I think you are raising a good question, and that is exactly what is the responsibility of the U.S. EPA in implementing and taking steps in the Bioterrorism Act of 2002 after we get the vulnerability assessments.

I don't read the statute as saying that EPA has a specific authority to follow up.

Mr. STUPAK. Well, when you look at the Presidential Decision Directive 63, issued in May 1998, it designated the EPA as the lead agency for assuring the protection of the Nation's water infrastructure. And so that was back in 1998, even before we had 9/11.

And then the Bioterrorism Act also makes that a requirement. You are the lead agency.

Mr. GRUMBLES. Congressman, a couple points: One is, we do have a broad authority under 1431, as you and your staff know full well, that if there is an imminent and substantial endangerment to public health, then under that provision—which has been in the statute for a long time prior to the 9/11 incident or the Bioterrorism Act of 2002—we will use our enforcement discretion and exercise that. We haven't done that to date.

With respect to the Presidential Directive 63, I mean, we do take seriously, and we continue to take seriously under the Presidential directives that have come after the Bioterrorism Act of 2002, our responsibilities to carry out the act and also to coordinate and do increased surveillance and monitoring.

Mr. STUPAK. Well, you know, we have got the Safe Drinking Water Act, we have Presidential Directive 63, we have the Bioterrorism Act of 2002. Is there some authority you want that would make it clearer for you your responsibility that you are the lead agency to protect the Nation's utilities and your water infrastructure in this country?

Is there some other authority you need or are looking for?

Mr. GRUMBLES. I think that it remains an open question as to whether or not Congress needs to revise the statute to provide us additional authority. I think we have our focus right now——

Mr. STUPAK. Well, we think we have given you three types of authority: Directive 63, the Safe Drinking Water Act, and the Bioterrorism Act.

Now, we are the guys who write this stuff, men and women who write it, so—but from your point of view, since you are supposed to be responsible for carrying it out, you tell us, are you missing some authority? Is someone telling you, Geez, that is a nice suggestion that I should do this to make sure the safe drinking water in Los Angeles is safer, but you know what, EPA, you don't have the authority.

Has anyone ever told you guys that?

Mr. GRUMBLES. I think, as I—I did want to emphasize in the statement, Congressman, Congress in reviewing the implementation of the Bioterrorism Act of 2002 should acknowledge and recognize that what that statute did, critically important and successful statute, was to set up a planning and vulnerability assessment framework, emergency response planning. We are carrying that out and implementing that, and so I am not here to seek additional regulatory or enforcement authorities. I know that our focus is on providing the tools and the training, the technical assistance to utilities to carry out their plans as they develop them.

Mr. STUPAK. Well, more than just vulnerability assessment. Again, go back to the letter I read to you earlier from the EPA Director Christy Todd Whitman again, once again, dated April 22, 2002.

And, Mr. Chairman, I ask this letter be made part of the record.

Mr. GILLMOR. Without objection.

[The information referred to follows:]

UNITED STATES ENVIRONMENTAL PROTECTION AGENCY
WASHINGTON, D.C. 20460

APR 22 2002

THE ADMINISTRATOR

The Honorable John D. Dingell
United States House of Representatives
Washington, D.C. 20515

Dear Congressman Dingell:

Thank you for your letter of February 26, 2002, expressing your concerns about the vulnerability of our Nation's public water systems to attack or other intentional acts. I want you to know that I am very committed to protecting our Nation's drinking water systems. The Environmental Protection Agency (EPA) is playing a significant role in working with State governments and local utilities to protect drinking water supplies. We have already begun working with States and local utilities to assess this vulnerability. Our efforts will ensure that utilities have developed a comprehensive assessment of these vulnerabilities and emergency operations plans using the most current methods and technologies.

I appreciate your support of our efforts to train water systems employees to conduct vulnerability assessments. I share your view on the importance of these assessments in assisting drinking water utilities mitigate their vulnerabilities. In response to your request, I am providing the following information to update you on EPA's efforts in this area.

On October 5, 2001, EPA established the Water Protection Task Force. This task force is charged with helping Federal, State and local partners expand and implement tools to safeguard the Nation's drinking water supply and wastewater treatment systems from terrorist attack. We have made considerable progress in our efforts to provide public water systems with the knowledge and tools to minimize the consequences of a terrorist attack.

The vulnerability assessment methodology for large drinking water systems developed by the Sandia National Laboratories has been tested at several major utilities (through an agreement with EPA). The tool lays out a process for utilities to assess vulnerabilities from attack on physical infrastructure, from contamination, from cyber attacks, and the failure of related systems on which the drinking water utilities depend (e.g., electrical energy supply). Upon completion of the assessments, utilities will have a risk-based, prioritized list of mitigation actions to consider.

EPA has trained thousands of utility operators and managers on the vulnerability assessment tool. We are also developing a "train-the-trainer" program that will cover vulnerability assessments and effective security measures. Together, these efforts will serve to develop a large body of expertise within the utility community and supporting science and engineering contractor community to perform the vulnerability assessments.

EPA is working with the Centers for Disease Control, the Department of Defense, the Food and Drug Administration, and the Federal Bureau of Investigation to improve information on drinking water contaminants including detection and treatment capabilities. This information will support vulnerability assessments and emergency response activities.

EPA has received Fiscal Year 2002 supplemental funds to support the conduct of vulnerability assessments and related security work. We will be using these funds to support public drinking water systems of all sizes with the goal of identifying vulnerabilities and putting in place appropriate measures to minimize the consequences of a terrorist attack.

Your letter also requests information on a number of related topics. Please find enclosed with this letter an attachment with answers to each of your questions.

I appreciate your support of our efforts at EPA. I want to thank you for writing me to highlight your concerns and for your continued interest in the protection of the Nation's drinking water. Should you need additional information or have further questions, please contact me or your staff may call Steven Kinberg, Office of Congressional and Intergovernmental Relations, at (202) 564-5037.

Sincerely yours,

Christine Todd Whitman

ATTACHMENT

1. **Would the EPA support legislation granting the Agency additional authority to require the adoption of safer technologies or enhanced security measures?**

EPA is continuing to explore this issue as it relates to drinking water utilities as well as other facilities that handle hazardous material. One of EPA's major responsibilities is to provide the training and tools to assist State and local governments in their decisionmaking processes. For example, the vulnerability assessment tool described earlier is designed to focus a drinking water utility on the risks associated with existing practices (e.g., use of gaseous chlorine) in the event of a terrorist act and examine the potential mitigation measures available to the utility.

2. **Has the EPA advised or made a formal recommendation to public water systems to replace gaseous chlorine with sodium hypochlorite or other safer alternatives such as ozonation or ultraviolet light and if not, why not?**

EPA has not made such a recommendation. There are important public health issues to consider relative to the appropriate balance of treatment effectiveness for both ongoing public health protection and protection from intentional contamination and risks associated with accidental or intentional releases of dangerous chemicals. One of the Agency's key concerns about recommending chlorine alternatives is the lack of treatment data on their effectiveness against certain contaminants that could be introduced into the drinking water system through malevolent acts. Alternatives such as ozone and ultra violet light do not provide the necessary disinfectant residual required for public health protection in the distribution system.

EPA has an active effort underway to meet with interested stakeholders and technical experts (e.g., environmental groups, drinking water utilities, industry associations, and security experts) to develop a guidance document on this topic.

3. **Has the Administration sought any new authority so the EPA or any other federal department can ensure that significant vulnerabilities detected at our drinking water utilities are corrected in advance of a terrorist or other intentional malicious act?**

Drinking water utilities have strong incentives to protect their ability to provide safe and reliable water to their customers. EPA believes that the vulnerability assessments will provide the systems with the information necessary to mitigate potential vulnerabilities. The Agency has sought appropriations for funding the conduct of vulnerability assessments. EPA is also providing assistance to systems to facilitate their use of the State Revolving Fund to pay for the correction of vulnerabilities identified in their assessments.

4. **Specifically, would you support adding language to the emergency powers authority (Section 1431) of the Safe Drinking Water Act that gives the EPA the authority to issue an order to a drinking water utility where the Agency has received information that a significant vulnerability exists that may present an imminent and substantial endangerment to human health or the environment?**

The existing language in Safe Drinking Water Act (SDWA), section 1431, does not provide broad general authority to require actions to address security concerns. Rather, section 1431 is limited to cases involving a contaminant that is present in or is likely to enter a public water system or underground source of drinking water and that may present an imminent and substantial endangerment to human health. The language contained in HR 3448 amending SDWA, section 1431, provides EPA with adequate authority to respond in situations involving significant vulnerability. More specifically, that language empowers the Agency to issue necessary orders where it has received information that there is a threatened or potential terrorist attack or other intentional act designed to disrupt the provision of safe drinking water or adversely impact the safety of drinking water and may present an imminent and substantial endangerment to human health.

5. **As vendors approach public water systems with new security devices, is EPA making any effort to inform water systems of the most effective technologies or devices and proper ways to employ such technologies or devices?**

EPA is in the process of investigating security-related detection, monitoring and treatment tools to identify currently available tools, and begin to fill gaps that currently exist. This effort is being conducted in partnership with the States and utility associations in order to determine which areas are a priority for additional work. These partnerships will also assist in determining who should take the lead in performing the work, and how best to disseminate the resulting information.

6. **Is EPA aware of any source of information available to public water systems that explain how a drinking water system should be configured or protected to be deemed secure? If not, is the EPA planning to develop a source of such information for different sized water systems?**

Since October 2001, EPA has distributed several notices to all water utilities suggesting some measures that all water systems can take to improve security. Because system design, operation and locational characteristics vary considerably from system to system, more in depth consideration on how to configure water systems to improve security requires system-by-system analysis. The vulnerability assessment methodology developed by Sandia National Laboratories for large drinking water systems is a process designed to help a water utility determine the specific steps the system should take. EPA will provide technical assistance to small and medium sized systems in the area of vulnerability assessments. EPA also plans to provide training on security measures as part of the strategy for supporting security improvements at small and medium systems.

7. **How many public water systems will receive grants in FY 2002 to conduct vulnerability assessments and what do you expect the average amount of each grant will be? Will the EPA, as a condition of receiving the federal grant, require criteria to insure that the vulnerability assessments are comprehensive and of high quality?**

Within the next few months, EPA expects to award approximately 400 grants to the "largest" water utilities. EPA considers a utility among the largest if it serves 100,000 or more persons. These utilities represent approximately 50 percent of the total population served by public water systems.

The Agency will provide approximately $53 million of Fiscal Year 2002 appropriations to the "largest" water utilities. The final number of systems receiving grants may vary, depending on the number of systems that may also serve 100,000 or more persons due to sale of water to a "consecutive system." These consecutive systems are being identified in cooperation with the States. If EPA awards 400 grants, the amount per system would be approximately $130,000.

EPA will stipulate in its grant guidance that the vulnerability assessments must contain specific elements that comprise a thorough assessment. Such an assessment contains the elements that exist within the Sandia National Laboratories' vulnerability assessment methodology, but other comparable processes may be used. EPA is also funding a train-the-trainer program that will help to spread the methodology to qualified consultants and utilities, and maintain a high degree of integrity in the assessment process.

In addition to the grants to the 400 largest systems, EPA will work with States, Tribes, and utility organizations to determine the best ways to meet small and medium drinking water system needs. EPA will use approximately $23 million of Fiscal Year 2002 appropriations to provide training, development and distribution of tools, and technical assistance to the small and medium systems.

Mr. STUPAK. The letter to Mr. Dingell. And again I go to page 2, the top paragraph. The language contained in H.R. 3448—that is the Bioterrorism Act—amending the Safe Drinking Water Act, section 1431, provides EPA with adequate authority to respond to situations involving significant vulnerability.

So according to the EPA Director, back then Christy Todd Whitman, you had significant authority to do what has to be done, and your job is really to make sure the public water supplies and distributions are secure from terrorist attack, more than just take assessments of utilities. You have a real responsibility here. And I am afraid the public, if they are watching this thing at all or hearing anything about this hearing, there is not a lot of assurance that the necessary security enhancements are being taken to make sure their water is safe.

Mr. GRUMBLES. Well, I would disagree with you respectfully, Congressman. There is no doubt that work needs to be done, and there is no doubt that EPA will exercise its existing authorities that it has in the Bioterrorism Act as well as the Safe Drinking Water Act. And it is also no doubt to us that there needs to be continued cooperative discussion, compliance assurance. Our top priority has been, Congressman, to ensure that the systems get in their vulnerability assessments and their emergency response plans, certify that they have done their emergency response plans, and that we work with the other Federal and State and local entities on workshops, tools and training, and update and improve their plans that they use—view them as living documents that need to be continuously improved.

Mr. STUPAK. With that answer, I take it you agreed with us that you have the authority; that you have done a bunch of assessments. But what I haven't heard you say in answer to my questions here today, you haven't taken any corrective action to make sure that these security enhancements are in fact in place. Your own internal document basically said the evaluations were based upon pre-9/11, which is basically vandals, criminals, and disgruntled employees, and because they didn't get any guidelines from you as to what we should be looking for post-9/11.

Mr. GILLMOR. The gentleman's time has expired. But the Chair would extend the gentleman's time long enough for me to ask the gentleman if he would yield to me.

Mr. STUPAK. I would be happy to yield to the chairman.

Mr. GILLMOR. I just want to point out as a factual matter the letter that you cited predated, as I understand it, the passage of the Bioterrorism Act, and whatever authority EPA may have, there was no specific authority that I am aware of in the Bioterrorism Act for them to take the action that you refer to. EPA may have it under other provisions, but I don't think under the Bioterrorism Act.

Mr. STUPAK. The language read in was, it was really—the question was, the reason why there was a letter between Christy Todd Whitman and Mr. Dingell was because they were asking about the existing language in the Safe Drinking Water Act, did it provide a broad enough general authority to require actions to address security concerns. But then they went into the language contained in 3448, which was the bioterrorism. And they felt that with the two of them, with both 3448, the Safe Drinking Water Act, Presidential Directive 63, they had more than enough authority to carry it out, not only just to ask for assessments, vulnerability assessments, but actually to take corrective action as they are the lead agency, as Directive 63 pointed out, to make sure that we have the assessments done properly post-9/11, corrective action be taken if necessary, and Congress was to, as the bioterrorism acts, appropriate moneys to make sure it is done. Of the $160 million that has been allocated, plus there was an emergency supplemental after 9/11 of $89, so about $240 million, $250 million, we have a lot of assessments that the expert says it isn't worth the paper it is written on and no corrective action since then.

Mr. GILLMOR. You and I are basically the spokesmen, Mr. Stupak, for dueling staff assessments, and the assessment that I am getting was that EPA asserted that authority before the passage of the act, but Congress didn't agree with that. But that is something we can get cleaned up at another time.

I want to thank the members who have participated in the hearing. I particularly want to thank Mr. Grumbles and Mr. Stephenson for your usual very helpful testimony, and the meeting stands adjourned.

[Whereupon, at 3:57 p.m., the subcommittee was adjourned.]

[Additional material submitted for the record follows:]

Mr. JOHN B. STEPHENSON
Director
Natural Resources & Environment
Government Accountability Office
441 G Street, NW
Washington DC, 20548

DEAR MR. STEPHENSON: This is to express our appreciation to you for testifying before the House Energy and Commerce Subcommittee on Environment & Hazardous Materials on September 30, 2004 for the hearing entitled Controlling Bioterror: Assessing Our Nation's Drinking Water Security.

Pursuant to the Chair's order, the hearing record remains open to allow Members to submit questions to witnesses. I would appreciate it if you could respond to these questions, and provide an electronic copy of your response no later than the close of business on Friday, October 29, 2004, in order to facilitate the printing of the hearing record. The electronic copy (in Word or WordPerfect format) can be e-mailed to Peter.Kielty@mail.house.gov.

Thank you again for your time and effort in preparing and delivering testimony before the Subcommittee.

Sincerely,

PAUL E. GILLMOR, *Chairman*
Subcommittee on Environment and hazardous Materials

Attachment

Question 1. According to an October 2003 report done for the Senate Committee on Environment and Public Works, GAO stated that security experts generally agree that decisions for allocating federal money for security improvements should be based primarily on (1) population density and (2) information contained in vulnerability assessments. Such efforts though could be complicated by Title IV's requirement that EPA develop protocols to protect from vulnerability assessments from disclosure to unauthorized individuals. As such, how do you square this recommendation with the requirements of the law?

Response. As authorized reviewers of Vulnerability Assessments (VAs), designated EPA officials may examine submitted VAs, and could use them in making funding decisions and recommendations without compromising the requirements of Title IV. As a practical matter, however, such funding decisions would be realistic only at an *aggregate* level (e.g., for making judgments about the future direction of research, the types of training and their target audiences, and other technical assistance). As we noted in our report, the use by EPA officials of VA information to make—and defend—decisions about allocation among *individual* recipients could indeed be complicated by Title IV's requirement to protect VAs from disclosure to unauthorized individuals. Experts also cited other complications that would complicate utility-specific allocation decisions based on VA information. For example, several noted that even if access to vulnerability assessments was available, using VAs would require a high degree of interpretation on *someone's* part, and it's not altogether clear how such judgments would be made among potential recipients.

Question 2. Based on your report, and recognizing the need for infrastructure funding, is it your opinion that some of this funding need for security enhancements should go through ratepayer increases, especially recognizing the current undervaluation of drinking water? Do you think it's reasonable to make local communities bear some of the costs in making these security upgrades?

Response. The degree to which the federal government supports utility efforts to improve security is a policy decision to be made by the Congress and the Administration. Our report sought advice on the most efficient ways to allocate and spend federal funds *should they be appropriated.* As a practical matter, many utilities are already financing at least some of their security upgrades by passing along the costs to their customers through rate increases. We would expect ratepayers to continue to shoulder much of these costs in the future. It is also worth noting that in responding to the question concerning desirable financing mechanisms, our experts voiced strong support for cost-sharing between the utility and the federal government, lending further weight to the notion that improved utility security is in large part a local responsibility.

Question 3. Your report recognized the physical assets of the distribution system as the single most important vulnerability of all system components. Recognizing the infrastructure needs of drinking water utilities and how the physical deterioration of pipes and transmission systems can lead to security vulnerabilities, do you agree with EPA that some SRF money helps improve security?

Response. As a financing mechanism, use of the SRF for security enhancements did not rank as high as a number of other mechanisms identified by our expert panel. Nonetheless, the majority of experts did site the Fund as either "very effective" or "somewhat effective" as an approach for distributing funds. Moreover, one would expect the SRF to be particularly appropriate in circumstances—as suggested in the question—in which addressing basic infrastructure needs ("physical deterioration of pipes and transmission systems") can, at the same time, address security-related concerns. The efficiency of this "dual use" concept has been widely accepted at EPA, among the experts on our panel, and elsewhere.

Question 4. In your opinion, and based on your report, can the three categories of security-enhancing activities: physical and technological improvements, education and training, and strengthening operational relationships; be achieved or strengthened without further congressional action? What is your assessment of how likely the utilities are to cooperate in this further action?

Response. There is little doubt that some of these security-enhancing activities would continue to take place without federal funds, and our report documents a number of utility initiatives to pursue some of them. At the same time, our work suggests, at least anecdotally, that the degree to which some of these enhancements are implemented will be a function of the level of federal support provided. For example, the experts overwhelmingly cited the use of real-time monitoring technologies as the single most important physical security enhancement that can be applied to drinking water facilities. However, many of the experts noted that smaller utilities would simply be unable to deploy these technologies without federal support. In addition, while regional collaboration is taking place within some states as our report noted (BASIC in the San Francisco area and MADIRT in North Carolina), there may be a need for federal attention to encourage collaboration in broader regions of the country.

Question 5. Recognizing the fact that the vulnerability assessment information is highly protected in order to protect sensitive information about each utility from those who may use the information to harm the utility, how, in your opinion and based on your study, is it possible to adopt security measures that both address vulnerabilities and mitigate the consequence of attack?

Response. The requirement for vulnerability assessments helps to ensure that each utility goes through the systematic process of identifying its vulnerabilities and, by extension, developing plans to address those vulnerabilities through the addition of preventive measures and response plans. In that sense, the secrecy imposed on vulnerability information by Title IV does not necessarily prevent utilities from adopting security measures that address vulnerabilities identified by their VAs.

Question 6. Recognizing that the primary mission of the Drinking Water SRF is to facilitate compliance with federal drinking water regulations and that this requirement alone makes the competition fierce and the funds scarce, do you believe that the drinking water SRF should be used as a main funding source for security enhancements at drinking water utilities?

Response. For the reasons cited in the question, we believe it would be inappropriate to rely on the SRF as a main source of funding for security enhancements, particularly if supplemental funding was not provided to the SRF specifically for this purpose. As noted in response to question #3, few of the experts on our panel supported the SRF as a primary source of funding for security enhancements, with some citing the competing demands already placed on the Fund for its primary purpose of funding basic infrastructure improvements.

Question 7. There is interest in the development and deployment of technologies that can detect contamination at the various stages of the community water system's intake valves, treatment plant, and delivery network. What is the status of these activities? How helpful will real time monitoring technologies, capable of providing near real time data for a wide array of potentially harmful water constituents, be in addressing security issues and why do you think this technology received the most support for federal funding than any other category?

Response. The development and deployment of advanced monitoring technologies are still in their early stages, according to EPA's 2004 "Water Security Research and Technical Action Plan." The Plan speaks, for example, of the continuing need to develop monitoring technologies for biological, chemical, and radiological contaminants and threats; and of the need to develop "drinking water early warning systems." The development and deployment of such technologies received the widest support of any single security-enhancing activity cited by our expert panel for the reasons cited in the question—they hold great promise in providing real-time data for a wide array of potentially harmful water contaminants. This capability is particularly crucial in the water distribution system: once a contaminant is introduced at this late stage, there is little protection between a potentially deadly contaminant and an unsuspecting public. In such a situation, time to alert unsuspecting consumers would be of critical importance, and a real-time monitoring capability may be the only option to provide that time.

Question 8. Recognizing the fact that drinking water distribution systems are so vulnerable due to their accessibility at so many points, do you envision the magnitude of the risk ever reaching a point where these systems could be fully and adequately protected? While water utilities have all assessed their vulnerabilities?

Response. It is hard to imagine a scenario in which all drinking water systems could be "fully and adequately protected." We believe, however, that well-conceived and properly funded security-enhancing strategies can help considerably to maximize deterrence against an attack; improve early detection should an attack take place; and improve response capabilities to help mitigate an attack's impacts. We also see value in encouraging utilities to revisit and upgrade vulnerability assessments over time; threats will likely change over time as will the strategies available to address deterrence, detection, and response.

U.S. ENVIRONMENTAL PROTECTION AGENCY
OFFICE OF THE INSPECTOR GENERAL

The Honorable PAUL E. GILLMOR
Chairman
Subcommittee on Environment and Hazardous Materials
Committee on Energy and Commerce
U.S. House of Representatives
Washington, DC 20515-6115

DEAR MR. CHAIRMAN: Enclosed are responses to questions for the record stemming from the September 30, 2004, hearing "Controlling Bioterror: Assessing Our Nation's Drinking Water Security." We appreciate the opportunity to comment on this important issue. If your staff should have any questions on these responses, please contact Eileen McMahon, Assistant Inspector General for Congressional and Public Liaison, at (202) 566-2391.

Sincerely,

NIKKI L. TINSLEY

Enclosure

RESPONSES TO QUESTIONS FROM CHAIRMAN GILLMOR

Question 1: In your report entitled "EPA Needs to Assess the Quality of Vulnerability Assessments Related to the Security of the Nation's Water Supply Report No. 2003-M-00013 Dated September 24, 2003," you cited that water systems did not consider the terrorist threat or distribution systems when undertaking their vulnerability assessments. Please clarify whether this conclusion was made before or after you were granted access to the vulnerability assessments. If the conclusions were drawn before you had access to the vulnerability assessments, would your conclusions change following your access?

Answer: We want to clarify that we did not state in our report that water utilities failed to consider terrorist threats or distribution systems when undertaking their vulnerability assessments. We reported in *EPA Needs to Assess the Quality of Vulnerability Assessments Related to the Security of the Nation's Water Supply* (Report No. 2003-M-00013), dated September 24, 2003, that "based on our interviews, we believe that vulnerability assessments submitted may emphasize traditional, less consequential, and less costly threats, such as vandalism or disgruntled employees. Therefore, vulnerability assessments *may not necessarily* address terrorist scenarios or the events of 9/11 that motivated passage of the Bioterrorism Act." (emphasis added) We based our conclusions on interviews with water security experts, EPA officials, and water utility personnel we talked with prior to gaining access to the vulnerability assessments. While the Act prohibits us from publicly discussing the information we obtained from the vulnerability assessments, the statements contained in our report remain valid.

Question 2: In your report entitled "EPA Needs to Assess the Quality of Vulnerability Assessments Related to the Security of the Nation's Water Supply Report No. 2003-M-00013 Dated September 24, 2003," you stated that neither the Bioterrorism Act nor EPA identified a minimum threat level against which water utilities should assess their vulnerabilities. However, this statement did not take into account that baseline threat information for vulnerability assessments of community water systems was the topic of an extensive stakeholder meeting where a wide variety of members from the water industry, including large systems, utilities, municipalities, and rural systems were represented. The consensus at the meeting was that the design basis threat selection should be left to individual utilities to account for the uniqueness of each water system while incorporating the threat information gained from local FBI offices and other security experts. How then do you suggest a federal standardized threat level in light of this evaluation, recognizing the inherent differences in community water systems nationwide?

Answer: As we reported, "neither the Bioterrorism Act nor EPA identified a minimum threat level against which water utilities should assess their vulnerabilities." Water security experts, including staff from Sandia National Laboratory (the contractor EPA used to develop one of the vulnerability assessment methodologies), support our conclusion that EPA should have set a minimum threat level against which utilities needed to assess their vulnerabilities. According to Sandia officials, EPA's practice of not setting minimum security measures left threat determinations open to interpretation, and thus inconsistent application of the vulnerability assessment methodology. For example, one water security expert contracted to conduct several large utility assessments said that, even after vulnerability assessment training conducted subsequent to the terrorist attacks on 9/11, water utilities tended to focus on vandals, criminals, and disgruntled employees.

Furthermore, in our report, *Survey Results on Information Used by Water Utilities to Conduct Vulnerability Assessments* (Report No. 2004-M-0001), dated January 20, 2004, state and local auditors found that 10 of the 16 water utilities utilized the Federal Bureau of Investigation (FBI) as a source of threat information, and only 3 of the utilities found FBI's threat information useful.

While we agree about the uniqueness of the vulnerabilities of each water system, even if EPA required utilities to assess threats at a standardized level, the utilities still had the flexibility to decide whether or how to protect against any vulnerability identified.

EPA's actions subsequent to the issuance of our report support our conclusion that EPA should have set a standardized threat level even in the face of unique utility characteristics. First, during an April 2004 interview with our team, an EPA official described the Agency's plans to conduct 60 threat scenario-driven emergency response field exercises across the country including training on "model emergency response plans" for utility consideration. Second, during an April 2004 meeting, a senior official from EPA's Water Security Division described the Agency's initiative to identify best security practices "since the water industry has very little standards for security." EPA based its initiative to develop minimum guidance on security enhancements on a utility's size (e.g., fence height, the need for intrusion alarms) and EPA will vary guidance for rural/small water systems since they face different security issues. Moreover, regional EPA staff with access to the vulnerability assessments believe that utilities still have not made the cultural leap to considering terrorist scenarios rather than focusing on fencing and lighting as response mechanisms. Finally, EPA formed a Water Security Working Group charged with: (1) identifying, compiling, and characterizing best security practices and policies for drinking water utilities; (2) considering mechanisms to provide recognition and incentives to implement them; and (3) considering mechanisms to measure the extent of implementation of these best security practices and policies.

○